Innovation in Clinical Trial Methodologies

Innovation in Clinical Trial Methodologies

Lessons Learned during the Corona Pandemic

Edited by

Peter Schüler, MD

Senior Vice President
Drug Development Services Neurosciences
ICON, Langen, Germany

ACADEMIC PRESS

An imprint of Elsevier

ELSEVIER

Academic Press is an imprint of Elsevier
125 London Wall, London EC2Y 5AS, United Kingdom
525 B Street, Suite 1650, San Diego, CA 92101, United States
50 Hampshire Street, 5th Floor, Cambridge, MA 02139, United States
The Boulevard, Langford Lane, Kidlington, Oxford OX5 1GB, United Kingdom

Notices
Knowledge and best practice in this field are constantly changing. As new research and experience broaden our understanding, changes in research methods, professional practices, or medical treatment may become necessary.

Practitioners and researchers must always rely on their own experience and knowledge in evaluating and using any information, methods, compounds, or experiments described herein. In using such information or methods they should be mindful of their own safety and the safety of others, including parties for whom they have a professional responsibility.

To the fullest extent of the law, neither the Publisher nor the authors, contributors, or editors, assume any liability for any injury and/or damage to persons or property as a matter of products liability, negligence or otherwise, or from any use or operation of any methods, products, instructions, or ideas contained in the material herein.

Library of Congress Cataloging-in-Publication Data
A catalog record for this book is available from the Library of Congress

British Library Cataloguing-in-Publication Data
A catalogue record for this book is available from the British Library

ISBN: 978-0-12-824490-6

For information on all Academic Press publications visit our website at
https://www.elsevier.com/books-and-journals

Publisher: Andre Gerhard Wolff
Acquisitions Editor: Erin Hill-Parks
Editorial Project Manager: Sam W. Young
Project Manager: Sreejith Viswanathan
Cover Designer: Alan Studholme

Typeset by TNQ Technologies

Contents

SECTION 3 Studies with less patient interaction

Contributors

Urs-Vito Albrecht, MD, PhD, MPH
Senior Scientist, Peter L. Reichertz Institute for Medical Informatics of TU Braunschweig and Hannover Medical School, Hannover, Germany

Frank M. Berger, MD
Head of Data Analytics, Global Clinical Operations, Boehringer Ingelheim, Ingelheim, Germany

Brendan M. Buckley, MD, DPhil, FRCPI
Chief Medical Officer, Teckro, Limerick, Ireland

Bill Byrom, BSc, PhD
Vice President of Product Strategy and Innovation, Signant Health, London, United Kingdom

Tim Clark, Dip. Stat., PhD
Vice President, Clinical Sciences, Drug Development Services, ICON Clinical Research GmbH, Munich, Germany

Simon Day, BSc, PhD
Director, Clinical Trials Consulting & Training Limited, Buckingham, United Kingdom

Melissa Hawking, BS
Senior Manager, Marketing, Symphony Clinical Research, Vernon Hills, IL, United States

Liselotte Hyveled, MscPharm, MMBS
Vice President, Research and Development, Novo Nordisk, Bagsvaerd, Denmark

Peter L. Kolominsky-Rabas, MD, PhD, MBA
Head, Interdisciplinary Centre for Health Technology Assessment (HTA) and Public Health, Friedrich-Alexander-University, Erlangen, Germany

Wayne Kubick
Chief Technology Officer HL7 International, Ann Arbor, MI, United States

Hans Lehrach
Alacris Theranostics GmbH, Berlin, Germany; Max Planck Institute for Molecular Genetics, Berlin, Germany; Dahlem Centre for Genome Research and Medical Systems Biology, Berlin, Germany; Southern University of Science and Technology, Shenzhen, Guangdong, China

William C. Maier, MPH, PhD
Vice President, Commercialisation and Outcomes, ICON Plc, London, United Kingdom

Fareed Mehlem
Medidata, a Dassault Systemes company, New York, NY, United States

Gareth Milborrow
Doctor, SVP Data & Applied Analytics IT ICON Clinical Research, Southampton, United Kingdom

Henrik Nakskov
Director Management CIMS 2015 Aps, Charlottenlund, Danmark

Lesley A. Ogilvie
Alacris Theranostics GmbH, Berlin, Germany; Max Planck Institute for Molecular Genetics, Berlin, Germany

Miguel Orri
AlgaeVir Sdn Bhd, Petaling Jaya, Selangor, Malaysia

John Reites
President, THREAD, Cary, NC, United States

Wolfgang Renz
Assistant Professor of Surgery, McGill University, Montreal, Quebec, Canada; Senior Lecturer, Faculty of Management, Technology and Law, Sankt Gallen University, Sankt Gallen, Switzerland

Damian T. Rieke
Charité Comprehensive Cancer Centre, Charité – Universitätsmedizin Berlin, Berlin, Germany; Berlin Institute of Health, Berlin, Germany

Peter Schüler, MD
Senior Vice President, Drug Development Services Neurosciences, ICON, Langen, Germany

Foreword

We live in unusual times. The Coronavirus Disease 2019 (COVID-19) has transformed the lives of our families and our communities in unprecedented ways, forcing us to adapt and shape a new future, one day at a time. While this new normal can be unsettling, it opens the doors to doing old things in new ways.

At the Drug Information Association (DIA), a nonprofit, individual membership organization committed to accelerating the delivery of therapies to patients worldwide, founded 56 years ago by visionary leaders during another health crisis—Thalidomide—we recognize the need for continuous evolution across the drug development continuum.

The landscape of clinical trial management has been evolving over the past several years given the development of innovative technologies, but accelerating and adapting clinical trial design has never been more important than at a time like the present, when we urgently need therapies for COVID-19. This pandemic has also emphasized the need to treat patients in new ways—a need to not just drive patient engagement in healthcare product development, but to create opportunities for technology to play a greater role via the use of smartphones, IoT, wearables, robotics, remote monitoring, big data integration, and more.

Clinical trial operations across the world have slowed down due to COVID-19. The effects of the pandemic on clinical trials are many: new trials are being put on hold and recruitment of patients is being suspended, due to a shortage of hospital staff available to manage clinical trials in the midst of healthcare systems being overrun by COVID-19 patients. Patients enrolled in the trial may also get infected themselves, need to drop out, or self-isolate, further affecting ongoing clinical trials. Digital health, described by the United States Food and Drug Administration (FDA), that includes "mobile health (mHealth), health information technology (IT), wearable devices, telehealth and telemedicine, and personalized medicine," provides a potential solution.

During the pandemic, virtual, hybrid, or decentralized clinical trials represent a logical way forward. However, the innumerable benefits of virtual clinical trials indicate that they may be the paradigm of the future. Virtual clinical trials utilize technology such as electronic monitoring devices and mobile applications. In current times, virtual visits and remote patient monitoring provide increased convenience and mitigate exposure risks associated with the pandemic. In the long term, virtual clinical trials enable sponsors to reach a broader patient population and have been shown to have higher patient recruitment rates, increased patient diversity, broader geographical representation, better compliance, and reduced dropout rates, all the while being faster and more cost-effective than traditional clinical trials [1]. The use of technology like wearable devices and smartphone applications enables collection of vast amounts of data, not always available in the clinical setting, in real time. While this can result in further challenges such as patient data privacy concerns, innovative technologies like blockchain can help.

Innovation in Clinical Trial Methodologies—lessons learned during the pandemic take us through the complex interplay of design, patient engagement, application of technology, and use of data to support decision-making and predictive outcomes.

Barbara Lopez Kunz, CEO, DIA Global
barbara.lopez.kunz@diahome.org;
+(202) 601-8901
DIA Global Center
21 Dupont Circle NW, Suite 300
Washington, DC 20036
United States

REFERENCE

[1] Ali Z, Zibert JR, Thomsen SF. Virtual clinical trials: perspectives in dermatology. Dermatology 2020. https://doi.org/10.1159/000506418.

Introduction

Is the Covid pandemic an accelerator for digitalization in our industry?

Peter Schüler, MD

Senior Vice President, Drug Development Services Neurosciences, ICON, Langen, Germany

The current digital revolution is the fourth media revolution of mankind after the invention of language (in the darkness of ancient times), writing (about 5000 years ago in the Near East), and book printing (over 500 years ago in the German Rhineland, invented by Gutenberg).

That last invention of book printing has many analogies to our current situation, since it was also stimulated by a pandemic:

Between 1347 and 1353, the Black Death killed a third of the European population. According to the thesis of the American scientist David Herlihy, the plague decimated a number of writers who were usually monks and had until then been tasked with copying scriptures. Due to their lack, the development of technical devices was stimulated which made the human labor less necessary—for example, for the printing of books.

According to other historians, the plague also left large collections of unused clothing from the deceased in Europe; these textiles, which were available free of charge, were extremely practical for the production of large-scale paper.

How fundamentally our thinking and behavior is conditioned by new media can also be understood by Gutenberg's invention. After 1450 it was the machine-reproduced book that reformatted man: universities and large libraries were founded; grammar books standardized the languages. Leonardo da Vinci thought about technical devices such as flying machines and hydraulic presses. The first pocket watch and the first globe were built and Columbus discovered America. The reformers came and peasant uprisings shook the old hierarchy of power [1].

What does that tell us about future drug development methodologies in a world after Corona?

We already observe a clear trend toward less labor-intense research: use of digital databases will continue to substitute "real" patients. The selection of sites and patients will follow an evidence-based selection algorithm. Telemedicine will replace at least in parts patient travel to the sites, and other outcome measures will be collected "naturally" through patient's wearables without any travel or assessment needs.

Innovation in Clinical Trial Methodologies. https://doi.org/10.1016/B978-0-12-824490-6.00015-3

The term "monitoring" will much more be linked to computer monitors, with future Clinical Research Associates (CRAs) checking data-flows and data-plausibility instead of reviewing paper documents on site.

More than ever before, we all need to "re-engineer" our daily work, similar to what the successful people in the Renaissance did. Otherwise our industry may disappear like one of the manuscript writers.

In the next section, Wolfgang Renz will throw a high level view on these innovative tools, while all subsequent sections will describe any such new technology or concept in more detail.

Reference

[1]   Hans-Jörg Künast and Wolfgang Reinhard. Spektrum Wiss 1993;6.

Best practices for streamlining development

Wolfgang Renz[1,2]

[1]*Assistant Professor of Surgery, McGill University, Montreal, Quebec, Canada;* [2]*Senior Lecturer, Faculty of Management, Technology and Law, Sankt Gallen University, Sankt Gallen, Switzerland*

The need

The current approach to bring a drug to market includes 6—9 years of clinical development and additional costs related to increased regulatory requirements, cost-effectiveness studies, high recruitment costs, and accounting for subpopulation needs in study design [1,2].

The evolution of technology in domestic and international settings requires diversity, necessity, and technology innovation [3]. To deliver true, sustainable innovation to clinical trials, these three elements should be assessed across the field of burgeoning products and processes.

With the convergence of mobile and electronic health, social media, and big data, the current model for developing novel medicines sees important changes: adjacencies, such as point-of-care devices, telemedicine, and portable and wearable technologies enable manufacturers to accelerate the development of innovative medicines. The phase II and III randomized control trials must be contextualized in the framework of real-world medication outcomes and thus must leverage from existing investments in health technologies, including social media, health informatics, telemedicine, biomedical innovations, mobile applications, and wearable monitors.

Still, the diversity of ideas will only lead to successful technologies if these idea makers and supporters include the perspective of future patients. If sponsors ask participants to include a technology model into their everyday life a framework for acceptance of the technology model will be required to be understood by the process designers of the trial. Low health and electronic literacy issues will continue to pose challenges if the technologies are designed with the user experience at front of mind. Additional failure rates, variable use case scenarios, and infrastructure needs will eliminate potential participants in trials unless the introduction of the technology is done with appropriate research on the population's diversity.

Patient privacy concerns have certainly contributed to the stunted growth in the field. The newest Cloud and health technology platforms in the consumer face have worked to overcome these challenges. Leveraging the device- and consumer-facing companies in health, clinical trial sponsors should be able to take lessons from their success combined with the latest in health information protection systems to

Innovation in Clinical Trial Methodologies. https://doi.org/10.1016/B978-0-12-824490-6.00013-X

overcome these challenges. Still, as technology evolution is organic and also dependent on the acceptance of human users, trust, acceptability, and usability are of utmost importance.

The solution

With the conversion of mobile and electronic health technologies, pharmaceutical manufacturers are aiming to incorporate "smart" devices into clinical development. New technologies and innovations provide the ability to screen, monitor, and communicate with study participants on an ongoing basis, abilities that will, in practice, decrease cycle times and reduce recruitment sizes if participants are more likely to stay engaged. Telemedicine is used to conduct remote trial visits, allowing for a more centralized approach to patient and participant engagement and follow-up. Collection of a new set of patient reported outcomes using accelerometer technologies that can assess calorie expenditure, sleep duration and quality, and activity are leveraged by trial sponsors and healthcare providers alike in their search for tools to analyze patient health and improvement. Innovative technologies using video, short messaging services (SMSs) or text, wearable monitors, and simple application programming interfaces to allow databases to speak to one another have already been pioneered in the consumer health sphere and are now poised to impact and drastically change the ways in which clinical trials are planned, conducted, evaluated, and sponsored.

The repurposing of new software and hardware technologies toward healthcare as well as the discrete design of electronic health (ehealth) and mobile health (mhealth) technologies for healthcare has provided for a new framework to envision more efficient clinical trials. These technologies (Table 2.1) and their actual or potential impact on clinical trials are examined here as are other nontechnology innovations and more distant future space for visionary clinical trials.

Social media

The ubiquitous nature of social media via Twitter, Facebook, and patient platforms has already begun to transform how current patients interact with the brand and manufacturer of their prescription drug [4]. With the recent first draft commentary from the Food and Drug Administration (FDA) on the acceptable uses of pharmaceutical

Table 2.1 Construct of new technologies for clinical trial, yet used in the consumer space.

	Consumer space	Clinical trials
Social media	High	Low
Electronic health records	Medium	Low
Telehealth	Moderate	Low
Wearable technologies	High	Moderate

advertising and interaction with current and potential patients, this stands to continue growing [5]. In the clinical trials space, social media has been used and has the potential being used as a patient recruitment tool. Platforms, especially disease-specific community sites (e.g., PatientsLikeMe) and disease management applications, reduce cycle time for recruitment of patients away from traditional healthcare trial enrollment points. Acurian was a pioneer in this area through launch of its clinical trial awareness and patient recruitment social networking application in 2008 [6].

A year after its introduction, Acurian announced Click it Forward, and their Recruitment Manager generated 50% of their clinical trials patient referrals through social medial platforms and networks [7, 8].

Electronic health repositories

Algorithms and computer-assisted culling of electronic medical platforms and other hubs of patient data can provide more accurate prediction of clinical trial recruitment. Recent research has demonstrated the feasibility of using health plan and registry databases to select the fewest feasible recruitment sites [9]. Simple algorithms based on inclusion and exclusion criteria can provide trial sponsors with identification of the most populous sites with the most number of eligible patients [10]. Preliminary research has also suggested that electronic health records (EHRs) patient indices may be predictive of a trial's ability to fulfill sample size and should be included as a preliminary check in trial assessment planning [11]. For more details see section "Data mining for better protocols" by F Melhem.

Clinical trial managers seeking patients already use EHR platforms to perform data capture after using it for matching of clinical trials. More cooperation among data capture systems and clinical trial stakeholders will be essential to improving efficiency and reliability of these processes. The Clinical Data Interchange Standards Consortium is one example of a successful consortium that seeks to enable higher levels of interoperability among data capture and clinical trial stakeholders. Also see Section "Data standards against Data Overload" by W Kubrick. Joint partnerships between data scientists, electronic record providers, and the pharmaceutical research industry will further develop the efficiency of clinical trial matching and recruitment, thus reducing cycle times and outreach costs. Affordable home-based DNA testing provided by companies (e.g., 23andMe) have increased access to genetic profiles and disease risk information while also providing a social community for learning and exchanging familiar disease and genealogical experience.

Telehealth

Telehealth was defined by the Institute of Medicine in 1996 as the "use of electronic information and communications technologies to provide and support healthcare when distance separates participants" [12]. Almost two decades later, telehealth or telemedicine still holds tremendous potential for achieving value across healthcare and more specifically in clinical trials. Telehealth, in its multiple channels, including mobile applications, video-conferencing, telephone, and SMSs or text, has applications for patient recording, visits, and monitoring adherence.

Study visits, often requiring study reimbursement, clinician and study personnel time, and other overhead costs related to location servicing, may be mitigated through the inclusion of remote and electronic visits. Although the FDA has voiced preference for in-person study visits, telemedicine has been accepted across a number of specific disease areas for traditional healthcare; thus, the Pfizer trial paved the way for alternate frameworks that allow for appropriate protocols to be modified for telehealth visits [8].

As trial sponsors look to innovate collection of patient-reported outcomes, investment by sponsors into customized, password-protected mobile applications to collect data is both reliable and potentially cost-effective. Although some of these systems have only been evaluated in consumer and post-trial market settings, they hold high likelihood for potential integration into clinical trials [13] and are of growing use. More is said in the section by J Reites "Telemedicine replaces site visits."

In addition, social media applications through web and mobile-based platforms are also suitable for collection of views of patients in the trial setting on perceived effectiveness, health status, and quality of measures [14]. The strength of this approach is regulatory approval for patient-reported collection instruments, which is suggestive of a causal, pragmatic pathway.

Patient-reported outcomes and all other study data collected through electronic data capture, through such systems such as Phase Forward and Medidata, hold a key perspective for understanding how systems have been adopted by pharmaceutical and life sciences partners and have actually generated revenue [14]. Electronic data systems and their historical demonstrated value add through shareholder revenue generated by public offerings, and large buyouts will provide an analog for other clinical trial innovations.

Wearable technologies

The new advent of wearable fitness monitors powered by triaxial accelerometers has tremendous application for adding value to clinical trials. Laboratory-validated accelerometers are largely accepted for providing reliable activity and sleep data. Add-on technologies for the current wearable triaxial accelerometers include heartbeat monitoring and pulse oximetry, improving upon the field of wearable devices. Details can be found in the section "The use of new digital endpoints" by B Byrom.

SWOT analysis

Strengths

- More connected trial
- Faster cycle times
- More data collection
- Fewer clinic visits (e.g., completed from home setting)
- Sourcing patients through social media increases recruitment times

Weaknesses

- Patient confidentiality
- Cloud-based data storage
- Data validation issues

Opportunities

- Following participants in real-world setting
- Potential decrease in clinical trial development costs

Threads

- Regulatory acceptance of data
- Big reliance on technology, multiple technology platforms to be sourced and integrated into database
- New entrants into the clinical trial space (e.g., technology companies) who will have considerable data expertise with customers.

Take-home message

Convergence of mobile and electronic health (e.g., EHR) will enable the next wave of innovation in clinical trials. Clinical trials need to be adapted for health-seekers or information-seekers that will have considerable information about their disease and treatment options. Policy and regulatory concerns, namely around exchange of confidential information, need to get addressed.

References

[1] Kaitlin K. Deconstructing the drug development process: the new face of innovation. Clin Pharmacol Ther 2010;87(3).

[2] Getz KA, Wenger J, Campo RA, Seguine ES, Kaitin KI. Assessing the impact of protocol design changes on clinical trial performance. Am J Ther 2008;15:450—7.

[3] Basalla G. The evolution of technology. Cambridge: Cambridge University Press; 1989.

[4] Steele R. In: Mukhopadhyay SC, Postolache OA, editors. Utilizing social media, mobile devices and sensors for consumer health communication: a framework for categorizing emerging technologies and techniques, vol. 2. Springer Berlin Heidelberg; 2013. p. 233—49. https://doi.org/10.1007/978-3-642-32538-0_11.

[5] U.S. Department of Health and Human Services, Food and Drug Administration. Guidance for industry: fulfilling regulatory requirements for postmarketing submissions of interactive promotional media for prescription human and animal drugs and biologics. 2014.

[6] Acurian. Click it forward. Click it forward web site. 2014. https://ols15.acuriantrials.com/jsp/facebook/default.html. Updated 2008. Accessed 2/9.

[7] Connor S.Acurian generates over 50% of clinical trial patient referrals from proprietary relationships with online health networks & social media platforms; 2014. Acurian Generates over 50% of Clinical Trial Patient Referrals from Proprietary Relationships with Online Health Networks & Social Media Platforms Web site. http://www. businesswire.com/news/home/20100304006150/en/Acurian-Generates-50-Clinical-TrialPatient-Referrals#. Uvgv7D84EQk. Updated 2010. Accessed 2/9.

[8] Pfizer tests a concept that could modernize drug studies. NJ.com. 2014. www.nj.com/business/index.ssf/2012/01/pfizer_tests_a_concept_that_co.html. [Accessed 2 September 2014].

[9] Curtis JR, Wright NC, Xie F, et al. Use of health plan combined with registry data to predict clinical trial recruitment. Clin Trials 2014;11(1):96−101.

[10] Sumi E, Teramukai S, Yamamoto K, Satoh M, Yamanaka K, Yokode M. The correlation between the number of eligible patients in routine clinical practice and the low recruitment level in clinical trials: a retrospective study using electronic medical records. Trials 2013;14:426−6215.

[11] Schwamm LH. Telehealth: seven strategies to successfully implement disruptive technology and transform health care. Health Aff 2014;33(2):200−6.

[12] Schapranow M, Plattner H, Tosun C, Regenbrecht C. Mobile real-time analysis of patient data for advanced decision support in personalized medicine. In: The fifth international conference on eHealth, telemedicine, and social medicine; 2013. p. 129−36.

[13] Baldwin M, Spong A, Doward L, Gnanasakthy A. Patient-reported outcomes, patient-reported information. Patient: Patient-Centered Outcomes Res 2011;4(1):11−7.

[14] Akami. Case study - phase forward. 2014. http://www.akamai.com/html/customers/case_study_phase_forward.html. Updated 2013. [accessed 2/11].

Alternative study concepts requiring less patients

Use of historic control data

Tim Clark, Dip. Stat., PhD [1], **William C. Maier, MPH, PhD** [2]

[1]*Vice President, Clinical Sciences, Drug Development Services, ICON Clinical Research GmbH, Munich, Germany;* [2]*Vice President, Commercialisation and Outcomes, ICON Plc, London, United Kingdom*

The need

Small patient populations in rare diseases, pediatric populations, and the desire by patients and their caregivers to receive active therapy while on a clinical trial make it much more likely that these trials will need to incorporate historical data into the design.

Rare disease clinical trials have several further challenges that make them even more difficult to conduct than clinical trials for more common diseases [1].

The solution

The inclusion or borrowing of historical data in the analysis of the planned trial could improve the precision of the estimates, thereby increasing the statistical power of the statistical test and reducing the sample sizes [2]. Historical data can either fully or partially replace the control group and be obtained from published clinical trials and/or from medical charts. The former can be problematic, as it is often difficult, if not impossible, to establish if the patients in the published trials are comparable to those to be included in the planned study. Less problematic are patient-level data from medical charts, particularly those obtained from a large center that regularly conducts clinical trials. In this case, it is often possible for participants in a recently completed study to form the external control group for the planned study.

The historical data must be chosen carefully using prespecified criteria. The most commonly used criteria for assessing the comparability of the historical and the planned trials are those proposed by Pocock [3]. In order for the comparison between the active and external (historical) control arms to be considered valid, the two populations should be exchangeable with one another with regard to the following: 1. Eligibility criteria, 2. Patient characteristics/confounders, 3. Mode of treatment, 4. Outcome measure, 5. Time period, 6. Clinical setting.

If the two populations are not perfectly exchangeable, which is commonly the case, then the comparison is potentially confounded (biased). Several approaches, based on frequentist and Bayesian methodologies, have been developed to limit

Innovation in Clinical Trial Methodologies. https://doi.org/10.1016/B978-0-12-824490-6.00009-8

the potential bias, including selecting a subset of controls to match the population eventually recruited in the planned trial or down weighting the historical controls versus concurrent controls in case of discordance [4].

Frequentist approach

- Individual patient-level data obtained from an external source (e.g., medical charts) are used to form a control group.
- Eligibility criteria for the planned study and the external control group are as close as possible.
- Approaches such as propensity score are used to balance covariates (the characteristics of participants) between the study and external control groups. The aim is to create a set of matched patients consisting of at least one participant in the study group and one in the control group with similar propensity scores.
- Event rates, mean effect, etc., in the study population are compared to those observed in the external patient population.
- Frequentist statistics evaluate the probability of the evidence given a hypothesis. It calculates the probability of an event in the long run of the experiment (i.e., the experiment is repeated under the same conditions to obtain the outcome). The frequentist analysis would give a point estimate with standard error, confidence interval, and *P*-value.

Bayesian approach

- Estimates of expected event rates, mean effect, etc., based on external subject-level data.
- These estimates are used to generate a prior probability distribution representing the likely values for the parameter of interest. This is then combined with the observed data from the clinical trial to update the belief about the parameter of interest. This is known as the posterior distribution, which is essentially a weighted average of the prior and the observed data.
- There are several options available for specifying a prior based on external subject-level data. The historical data can be discounted (down weighted) based on the degree of similarity between external and study patients. These approaches can be augmented by expert opinion and/or other sources of historical data.
- In contrast with frequentist methods, Bayesian inference does not rely on the concept of infinitely repeating an experiment. Instead, it starts with the prior belief and then updates this when data from the study become available.
- The Bayesian analysis generates the mean and standard deviation of the posterior distribution together with the credible interval, which is the probability that the true (unknown) estimate would lie within the interval, given the evidence provided by the observed data.
- By borrowing from appropriate prior information, the same decision might be reached with a smaller sample size and/or fewer patients exposed to a placebo control arm.

More detailed information on the use of both Bayesian and Frequentist methods in the selection and analysis of historical control populations along with examples of the successful use of historical controls in drug approval is given in Lim et al. "Minimizing Patient Burden through the Use of Historical Subject-Level Data in Innovative Confirmatory Clinical Trials: Review of Methods and Opportunities" (https://www.ncbi.nlm.nih.gov/pubmed/29909645) [4].

An additional reference is the course presentation by Marc Walton of the FDA on "Historical Controls for Clinical Trials—Contemplation on Use in Drug Development" (https://events-support.com/Documents/Walton_Marc.pdf). This presentation gives additional detail of how historical (external) controls may be beneficial and examples from regulatory applications prior to 2012.

In the following sections, we explain the main differences between the two approaches, provide an overview of the regulatory guidance, and give some pointers on how to incorporate historical data into the study design.

How to incorporate historical data

The process should begin with identifying the sources of historical data, and all the important characteristics that need to be balanced between the treatment and control groups [5]. Exchangeability should be maximized by the use of common eligibility criteria for both the treatment and the historical groups. Differences are nonetheless inevitable, so these should be identified upfront in order to decide how they should be handled in the statistical analysis.

Evaluating and sourcing external data for historical controls

1. Prospectively establish a search plan for identifying and selecting the historical data, which could be subject-level control data and/or historical data from other sources. This could include information from published literature.
2. Evaluate historical data to determine suitability on following criteria:
    (a) Availability of relevant outcomes and other necessary data elements
    (b) Similarity to the treated group in all respects including disease severity, duration of illness, prior treatments, and any other aspects of the disease that could affect the measurement and timing of outcomes
    (c) Period of data collection relative to timing of clinical trial
    (d) Number of patients
    (e) Duration of patient experience
    (f) Ability to conduct individual-level analysis
    (g) Data licensing process and use restrictions
    (h) Cost of data access
3. If limited data exists then expert elicitation of the prior distribution could be considered. Expert elicitation is a formal means of extracting information from experts to either replace or augment external data.

The decision to take a Bayesian or frequentist approach to borrowing historical information is generally influenced by the availability of subject-level information to compare with the active trial arm and the preference of regulatory agencies in a specific disease area or application. International regulatory agencies have normally favored frequentist approaches when a historical control is completely substituted for a concurrent control arm in a phase 3 or registration study. Bayesian approaches have tended to be confined to early phase studies or extrapolating across different populations, e.g., from adults to pediatrics.

Key concepts

(A) Borrowing Information in a Bayesian Approach

A Bayesian analysis requires specification of a prior probability distribution reflecting what is currently known about the parameter of interest, for instance, the response rate [6]. The prior distribution is combined with the data observed in the trial (likelihood) to form an updated (posterior) distribution. This is the probability of the response rate given the data and can answer a question such as, "For Drug X, what is the probability that the response rate is more than 80%?" [7] This is depicted in Fig. 3.1 [8].

This posterior is a weighted average of the information in the prior and in the observed data, weighted by their relative precisions, which are ultimately associated with sample sizes. Thus, a natural way of including historical data in the analysis of a prospectively planned trial is by using it to construct a prior distribution for the control response rate. When a single historical study is available, the most direct way to

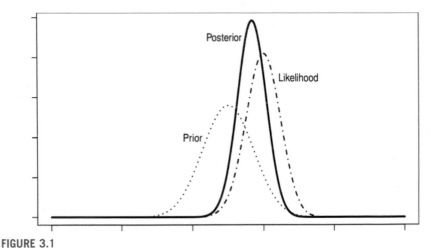

FIGURE 3.1

Historically known distribution of the parameter of interest (=Prior), data observed in the trial (=Likelihood), and the updated (=Posterior) distribution.

do this is to use the sampling distribution of the response rate in the historical trial as the prior for control response rate in the current trial. This turns out to be equivalent to pooling the historical and current trial data. The approach can be extended to multiple historical studies by pooling the historical studies and treating them as a single large historical trial, being careful to account for between-study variation. However, pooling historical and concurrent controls only seems justifiable under very specific and tightly controlled situations where it is reasonable to assume that the true underlying control rate in the population is the same in both historical and current settings.

Several other options are also available for specifying a prior based on historical data (see Table 3.1). These options reflect a range of different, and less stringent, assumptions about the relevance of the historical data and the relationship between the true control response rates in the populations represented by the historical (p_H) and concurrent controls (p), respectively.

All of these methods are associated with a level of subjectivity. In order to introduce greater objectivity into the amount of discounting or weighting applied to the historical data, a number of dynamic borrowing methods have been developed. These include hierarchical metaanalytic models (MAP priors), which assume that the historical and current response rates are exchangeable. The amount of borrowing depends on the between-trial heterogeneity estimated from the data. Large differences between concurrent and historical controls would result in very limited borrowing using this approach and vice versa. Power priors discount the historical data based on the observed difference between historical and concurrent controls. Finally, robust MAP priors assume that the historical and current response rates are exchangeable, but the prior is a mixture of a historical MAP and vague prior.

Table 3.1 Various methods of discounting or weighting the historical data in the estimation of the true control response rate.

	Assumption about differences between p_H and p
Equal	Due to sampling variation. Equivalent to pooling historical and concurrent controls.
Functional dependence	Explained by known covariates. Need to estimate covariate-response relationship (e.g., from historical data), which can be used to construct a prior distribution.
Equal but discounted	Assumes $p_H = p$, but discounts historical data by inflating variance of historical prior. Essentially reduces sample size on which prior is based. Amount of discounting can be based on expert opinion.
Biased	p is a biased version of p_H. Prior distribution for p constructed by combining the prior for p_H with the prior for the bias parameter, which is chosen to reflect the quality and extent of the historical data.
Exchangeable	p and p_H are similar, i.e., drawn from the same distribution with a variance parameter reflecting the heterogeneity between historical and concurrent control response rates.

The following publications provide a more detailed description of these methods with practical examples:

1. Rosmalen et al. *Including Historical Data in the Analysis of Clinical Trials: Is It Worth the Effort?* (https://pubmed.ncbi.nlm.nih.gov/28322129/) [9].
2. Lim J et al. *Minimizing Patient Burden through the Use of Historical Subject-Level Data in Innovative Confirmatory Clinical Trials*: https://pubmed.ncbi.nlm.nih.gov/29909645/) [4].
3. Spiegelhalter et al. *Bayesian Approaches to Clinical Trials and Health-Care Evaluation.* (https://onlinelibrary.wiley.com/doi/book/10.1002/0470092602) [10].

Finally, Laura Thompson provided examples in the following presentation (https://www.fda.gov/media/87358/download) of where the FDA believes Bayesian methods could be used in studying rare conditions in pediatric populations [11]. These include developing more realistic estimates of adverse event rates in small populations; drawing statistical strength from adult data to make decisions about device performance in pediatrics; and shortening trials through use of adaptive designs and predictive probability of trial success before all patients finish the trial.

(B) Adjusting for Differences in a Frequentist Approach

Patients in the external control may be different from those included in the active control population. Patient factors affecting both the selection of therapy and the probability of the treatment outcome have the ability to confound (bias) the results of comparison between the historical control and active population.

The differences may be reduced by several different strategies. These include restriction of the patient population entering the historical control, matching of historical control to increase similarity with patients in the active arm, and/or statistical methods to adjust for observed differences.

Propensity scores have been used as a way of adjusting for multiple patient factors in nonrandomized studies. A propensity score is the probability of treatment assignment conditional on observed baseline characteristics. The propensity score allows for the design and analysis of a nonrandomized study so that it mimics some of the particular characteristics of a randomized controlled trial. Additional detail on propensity score calculation (also see Table 3.2) and their use in nonrandomized studies is provided by Peter Austin in "An Introduction to Propensity Score Methods for Reducing the Effects of Confounding in Observational Studies" (https://www.ncbi.nlm.nih.gov/pmc/articles/PMC3144483/) [12].

Resources for use in development of external controls

1. Ongoing or newly established patient registries identified through literature review and other sources.
   (a) A major source of information on existing rare disease registries is provided by Orphanet. Orphanet provides an online searchable database for use in the identification of ongoing patient registries and other research projects. https://www.orpha.net/consor/cgi-bin/ResearchTrials.php?lng = EN

Table 3.2 Three methods propensity scores are used to reduce the influence of differences between the external and active control groups in nonrandomized studies.

Method	
Matching	Match individual patients on propensity scores or to create a similar distribution in both historical control and active treatment arm.
Stratification	Stratify based on propensity scores and develop weighted estimate of treatment effect.
Inverse probability of treatment weights (IPTW)	Weighting subjects by the inverse probability of treatment received creates a synthetic sample in which treatment assignment is independent of measured baseline covariates.
Covariate adjustment on propensity score (CAPS)	Use propensity score as a covariate in multivariate model to compare treatment effect.

    (b) The National Organisation of Rare Diseases also maintains a searchable database of resources for patient, physicians, and researchers in rare diseases, which may be used to identify organizations running patient registries. https://rarediseases.org/for-patients-and-families/information-res ources/rare-disease-information/

    (c) EURORDIS-Rare Diseases Europe is a unique, nonprofit alliance of 884 rare disease patient organizations from 72 countries that work together to improve the lives of the 30 million people living with a rare disease in Europe. There is a search engine to find information on rare diseases within the EURORDIS websites and websites of other rare disease organizations. https://www.eurordis.org/find-information-on-your-disease

    (d) In case established registries do not provide useful data, you can also establish a tailored registry prior to the planned trial (also see section "Patient-centric registries for population enrichment").

2. Healthcare databases derived from medical records

    (a) Electronic healthcare databases in Europe: descriptive analysis of characteristics and potential for use in medicines regulation [13].

        (i) https://bmjopen.bmj.com/content/8/9/e023090

3. Retrospective studies using nonelectronic medical records.

    (a) If none of above sources is available, you can perform such a data review prior to the planned trial.

Examples (Bayesian applications)

Benlysta (belimumab) approval

FDA-approved intravenous (IV) belimumab (BEL) in April 2019 for the treatment of children 5—17 years of age with active, seropositive systemic lupus erythematosus (SLE) receiving standard care (SOC). The approval was supported by a randomized trial that evaluated the efficacy, safety and pharmacokinetics (PK) of 10 mg/kg IV BEL versus placebo in 93 pediatric patients. Due to the rarity of the disease in children, a fully powered phase 3 pediatric study was not feasible. Determination of efficacy was therefore based on PK and efficacy results from the study, as well as extrapolation of the established efficacy of IV BEL from the two phase-3 adult studies. To provide more reliable efficacy estimates, FDA performed a post-hoc Bayesian analysis, which borrowed information from the phase-3 adult IV studies, under the assumption that outcomes would be similar in adults and pediatric subjects [14].

Phase-3 noninferiority study in a rare form of cancer

Based on the overall design and eligibility criteria for the planned study, therapeutic experts selected the historical data for inclusion in a metaanalysis. All data were generated from randomized, placebo controlled studies with time-to-event endpoints. The same statistical methodology (Cox regression) was used to calculate the Hazard ratio (HR) and 95% confidence interval (CI).

The sample size was calculated for a classical noninferiority study comparing test to control drug using fixed-margin (95%—95%), point-estimate, and Bayesian methods. The same noninferiority margins, control hazard rate, significance level, power and randomization ratios were used in all sample size calculations.

An informative, mixed prior for the control group and a noninformative prior for the test group were used in the Bayesian sample size determination. There was existing information on the likely performance of the control group, but limited or no information on the test group.

The Bayesian approach resulted in significantly lower sample sizes for the proposed study, approximately half that required for a classical noninferiority study.

Phase-3 equivalence study for a proposed biosimilar in patients with Diabetic Macular Edema (DME)

The primary endpoint was change from baseline in mean Best Corrected Visual Acuity (BCVA). Based on the proposed study design and eligibility criteria, we and other therapeutic experts selected historical data for inclusion in a metaanalysis to construct a prior distribution. Historical data used were from randomized, placebo controlled studies with change from baseline in mean BCVA as the primary endpoint. Mixed model repeated measures [MMR] were used to calculate the pooled treatment effect and 95% confidence interval (CI).

We took the historical data included in the metaanalysis results and constructed an informative prior for the control group and a noninformative prior for the test group. There was existing information on the likely performance of the control group, but limited or no information on the test group (Fig. 3.2).

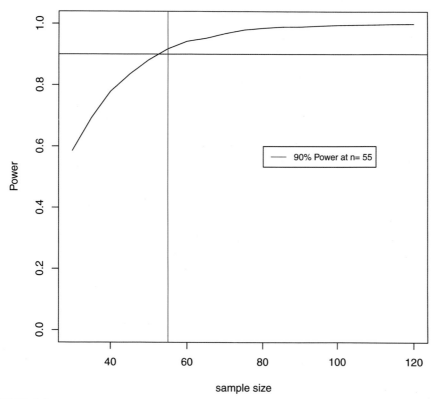

FIGURE 3.2

Sample sizes needed for different powers. Given the agreed power was to be 90%, alpha of 0.05 is reached at approximately N = 55 for reference, N = 110 for the biosimilar. Calculations were performed in R.

Examples (frequentist applications)

Brineura (cerliponase alfa) approval

Historical control information from an ongoing patient registry was used in the approval of Brineura (cerliponase alfa), the enzyme replacement therapy that helps treat CLN2 disease, a common form of Batten disease. Brineura was approved in April 2017 by the FDA to slow the loss of ability to walk or crawl (ambulation) in symptomatic pediatric patients 3 years of age and older with CLN2 disease [15].

The historical control cohort was identified from an ongoing patient registry and clinical information was collected from records and patient interviews. There were differences in patient characteristics between the clinical study and the historical control populations, namely age, genotype, and gender. Another limitation in the registry data was a lack of key outcome information. Efficacy in the clinical study was measured using a clinical rating scale developed specifically for CLN2 patients. Although the original instrument included four domains (Motor, Language, Visual, and Seizures), only two domains (Motor and Language) were measured in the historical cohort because accurate data could not be collected retrospectively for the other domains.

When the efficacy data were analyzed at 48 weeks, the findings were inconclusive. The FDA analyzed efficacy data at two additional time-points: 72 and 96 weeks. These analyses included: a "best matching" analysis based on the 96-week time-point; an ordinal analysis at 96 weeks that also included the 48- and 72-week time-points for repeated measures analysis. In the best match analysis patients were matched by baseline motor score, baseline age, and genotype. FDA concluded that despite the dissimilarities in patient characteristics, a distinct difference in motor function efficacy has been established for Brineura.

Exondys 51 (eteplirsen) approval

Exondys 51 (eteplirsen) injection was approved by the FDA in 2016 as the first drug approved to treat patients with Duchenne muscular dystrophy (DMD) [16]. The drug approval process involved the use of a natural history population as an external control. Initially the sponsor conducted a 24-week randomized placebo-controlled study, which did not meet its primary endpoint. The FDA advised the sponsor to conduct a new randomized, placebo-controlled trial. Many in the DMD community had strong reservations regarding the ethics and practicality of conducting another placebo-controlled trial of eteplirsen. Given the apparent difficulty of doing such a trial, FDA expressed willingness to consider an externally controlled trial.

The FDA advised the sponsor to identify external control groups and match them to patients in the original clinical trial. The sponsor identified two DMD patient registries as a source of external data, the "Italian DMD Registry" and the "Leuven Neuromuscular Reference Center" registry and conducted a post hoc comparison of the patients in the eteplirsen group with patients from the two external registries. The sponsor matched patients based on five factors: corticosteroid use at baseline; sufficient longitudinal data for Six Minute West Test (6MWT) available; age ≥ 7 years; genotype amenable to any exon skipping therapy; and genotype amenable to exon 51 skipping therapy. The primary clinical efficacy outcome measure was the 6MWT. Although this analysis failed to provide evidence of a clinical benefit of eteplirsen when compared to the external control group, the external control group information was used extensively in the medical review process.

The accelerated approval of Exondys 51 was ultimately based on the surrogate endpoint of dystrophin increase in skeletal muscle observed in some Exondys 51-treated patients. The FDA concluded that the data submitted by the sponsor demonstrated an increase in dystrophin production that was reasonably likely to predict clinical benefit in some patients with DMD who have a confirmed mutation of the dystrophin gene amenable to exon 51 skipping. Under the accelerated approval provisions, the FDA required the sponsor to conduct a clinical trial to assess whether Exondys 51 improved motor function of DMD patients with a confirmed mutation of the dystrophin gene amenable to exon 51 skipping.

Cystic fibrosis medication comparison with external control

The objective was to identify historical data from an ongoing Cystic Fibrosis (CF) registry for use as an external control in the comparative analysis of results from multiple clinical trials of lumacaftor/ivacaftor combination therapy. The primary objective was to assess the long-term safety of combined therapy. The estimated annual rate of decline in percent predicted FEV1 (ppFEV1) in treated patients was compared with that of a matched registry cohort. The long-term safety profile of lumacaftor/ivacaftor combination therapy was consistent with previous randomized clinical trials. Benefits continued to be observed with longer-term treatment and lumacaftor/ivacaftor was associated with a 42% slower rate of ppFEV1 decline than in matched registry controls. See additional details in the referenced publication below.

The analysis was judged successful and published in a peer-reviewed journal (Konstan et al. *Assessment of Safety and Efficacy of Long-Term Treatment with Combination Lumacaftor and Ivacaftor Therapy in Patients with Cystic Fibrosis Homozygous for the F508del-CFTR Mutation (PROGRESS): A Phase 3, Extension Study* [https://pubmed.ncbi.nlm.nih.gov/28011037/]) [17].

Applicable regulations

Table 3.3 provides a description and link to relevant regulatory guidance on use of historical controls in clinical trials from the EMA and the FDA.

Take-home message

1. Several examples give evidence that historical data can either fully or partially replace the control group.
2. The historical data must be chosen carefully using prespecified criteria. Trial data, registries, health records, and chart review are potential sources.
3. If the two populations are not perfectly exchangeable, which is commonly the case, then the comparison is potentially confounded (biased). Several approaches, based on frequentist and Bayesian methodologies, have been developed to limit the potential bias.
4. The decision to take a Bayesian or frequentist approach to borrowing historical information is generally influenced by the availability of subject-level information and the preference of regulatory agencies in a specific disease area or application.

Table 3.3 Available guidelines and regulation

ICH E10: Choice of control group and related issues in clinical trials	Describes the usefulness of such controls under certain scenarios. The guideline describes situations where appropriately and carefully chosen historical controls are more persuasive and potentially less biased	https://www.ema.europa.eu/en/documents/scientific-guideline/ich-e-10-choice-control-group-clinical-trials-step-5_en.pdf
FDA rare diseases: Natural history studies for drug Development—guidance for Industry	Defines natural history study and outlines applications of natural history data within rare disease drug development including use of historical controls.	https://www.fda.gov/media/122425/download
EMA guideline on clinical trials in small populations	States that "under exceptional circumstances historical controls with no concurrent control may be acceptable."	https://www.ema.europa.eu/en/documents/scientific-guideline/guideline-clinical-trials-small-populations_en.pdf
ICH E9: Statistical principles for clinical trials	States that "Bayesian and other approaches may be considered when the reasons for their use are clear and when the resulting conclusions are sufficiently robust."	https://www.ema.europa.eu/en/ich-e9-statistical-principles-clinical-trials
FDA guidance for the use of Bayesian statistics in medical device clinical trials	Provides guidance on statistical aspects of the design and analysis of clinical trials for medical devices that use Bayesian statistical methods.	https://www.fda.gov/regulatory-information/search-fda-guidance-documents/guidance-use-bayesian-statistics-medical-device-clinical-trials
Interacting with the FDA on complex innovative trial designs for drugs and biological products	Provides guidance to sponsors and applicants on interacting with the FDA on complex innovative trial design (CID) proposals, including Bayesian approaches, for drugs or biological products.	https://www.fda.gov/regulatory-information/search-fda-guidance-documents/interacting-fda-complex-innovative-trial-designs-drugs-and-biological-products

References

[1] Griggs RC, Batshaw M, Dunkle M, et al. Clinical research for rare disease: opportunities, challenges, and solutions. Mol Genet Metab 2009;96(1):20−6.

[2] Viele K, Berry S, Neuenschwander B, et al. Use of historical control data for assessing treatment effects in clinical trials. Pharm Stat 2014;13(1):41−54.

[3] Pocock SJ. The combination of randomized and historical controls in clinical trials. J Chronic Dis 1976;29(3):175−88.

[4] Lim J, et al. Minimizing patient burden through the use of historical subject-level data in innovative confirmatory clinical trials: review of methods and opportunities. Ther Innov Regul Sci 2018;52(5):546−59.

[5] Ghadessi M, Tang R, Zhou J, et al. A roadmap to using historical controls in clinical trials − by drug information association adaptive design scientific working group (DIA-ADSWG). Orphanet J Rare Dis 2020;15:69.

[6] Berry D. Bayesian clinical trials. Nat Rev Drug Discov 2006;5:27−36.

[7] Lee JJ, Chu CT. Bayesian clinical trials in action. Stat Med 2012;31(25):2955−72.

[8] Albert J. Bayesian computation with R. 2nd ed. 2009. New York, Springer.

[9] Rosmalen, et al. Including historical data in the analysis of clinical trials: Is it worth the effort? Stat Method Med Res 2018;27(10):3167−82.

[10] Spiegelhalter, et al. Bayesian approaches to clinical trials and health-care evaluation. Wiley; 2003. https://doi.org/10.1002/0470092602. Print ISBN:9780471499756 |Online ISBN:9780470092606.

[11] Source: Bayesian Methods for Making Inferences about Rare Diseases in Pediatric Populations, L.A. Thompson, Division of Biostatistics/OSB, CDRH, FDA - https://www.fda.gov/media/87358/download.

[12] Austin PC. An introduction to propensity score methods for reducing the effects of confounding in observational studies. Multivariate Behav Res 2011;46(3):399−424.

[13] Pacurariu A, Plueschke K, McGettigan P, et al. Electronic healthcare databases in Europe: descriptive analysis of characteristics and potential for use in medicines regulation. BMJ Open 2018;8. https://doi.org/10.1136/bmjopen-201. e023090.

[14] Benlysta® (belimumab). NDA/BLA multi-disciplinary review and evaluation. October 12, 2018.

[15] FDA summary review for Brineura. 2015. https://www.accessdata.fda.gov/drugsatfda_docs/nda/2017/761052Orig1s000SumR.pdf.

[16] FDA Summary Review for Exondys 51. 2016. https://www.fda.gov/news-events/press-announcements/fda-grants-accelerated-approval-first-drug-duchenne-muscular-dystrophy.

[17] Konstan MW, McKone EF, Moss RB, et al. Assessment of safety and efficacy of long-term treatment with combination lumacaftor and ivacaftor therapy in patients with cystic fibrosis homozygous for the F508del-CFTR mutation (PROGRESS): a phase 3, extension study. Lancet Respir Med 2017;5(2):107−18. https://doi.org/10.1016/S2213-2600(16)30427-1.

Adaptive and platform trials

Simon Day, BSc, PhD

Director, Clinical Trials Consulting & Training Limited, Buckingham, United Kingdom

The need

Clinical trials, as traditionally implemented, involve long periods of time where no-one knows what the results might be looking like. Although such an approach (the classic, double-blind trial) has advantages in terms of eliminating certain types of bias, it also means that if anything could—or ought to— be usefully modified to help the trial better answer the question it is intended to address, the sponsor will be unaware of this and so not in a position to make such changes. Only after the trial has finished will any deficiencies become known. This sometimes results in a trial being of little or no value. A simple example would be if it became apparent that the dose being tested was wrong; it would seem sensible to have the ability to change the dose as soon as that became apparent rather than not discover this fact until the trial had finished. Another example might be to allow, or disallow, a concomitant medication part way through a study. Particularly in the context of disallowing a co-medication, it is worth considering what action we might take if, early on in a trial, an untoward adverse interaction is observed between the treatment under investigation, and a sometimes-used co-medication. Clearly the trial cannot continue as it is, risking exposing future patients to a harmful combination of (individually unharmful) therapies. The trial could be stopped on safety grounds; a new trial might then be initiated with an exclusion criterion disallowing co-use of the other medication. This would clearly result in delay (and increased cost) to complete a further trial. It would also, in some sense, "waste" the information from patients in the first trial who were *not* taking the co-medication and who would be exactly the sort of patients recruited to the new trial. A compromise might be to start a new trial, but include these eligible patients from the first trial in some sort of meta-analysis with the second trial. A much neater and more efficient solution would be to continue the initial trial, but with appropriately restricted inclusion criteria. These are the beginnings of the ideas of adaptive designs: some feature(s) of the design can change, or be adapted, as the trial ensues. Having appropriate methodology (largely, but not exclusively, statistical tools) and the operational ability to make changes midway through a trial may avoid many lost opportunities.

In fact, numerous changes have been made to ongoing clinical trials for many years. Some have been good, some bad. The list below suggests some changes

Innovation in Clinical Trial Methodologies. https://doi.org/10.1016/B978-0-12-824490-6.00016-5

that do happen (from minor and common place, through to more substantial and on to extreme). We will see that some of these might fit into the term "adaptive design," even though they have been used well before the term was even thought of.

In an ongoing trial you may change:

- Inclusion criteria
- Sample size (increase or decrease)
- Endpoints (including the timing, the method of measuring these)
- Dose or dosing regimen
- Control or experimental treatment
- Primary objective

Some of these look common-place and are typically handled as protocol amendments. Some (the latter few on the list) seem very extreme.

The solution

So what classes as an "adaptive design" as opposed to a simple protocol amendment or something one would call "a bad idea"? Full consensus is lacking, but the CHMP Reflection Paper (CHMP/EWP/2459/02, October 2007) used this definition:

A study design is called "adaptive" if statistical methodology allows the modification of a design element (e.g., sample size, randomization ratio, number of treatment arms) at an interim analysis with full control of the Type I error.

This seems quite widely encompassing (any "modification of a design element") but restricts the definition only to cases where full control of Type I error can be assured. Control of Type I error is important, but the inclusion of this as a requirement within the definition may, in fact, be too restrictive.

FDA draft guidance ("Adaptive Design Clinical Trials for Drugs and Biologics") from February 2010 (and lasting through to September 2018, see below) stated the following:

"For the purposes of this guidance, an adaptive design clinical study is defined as a study that includes a prospectively planned opportunity for modification of one or more specified aspects of the study design and hypotheses based on analysis of data (usually interim data) from subjects in the study."

It then went on to add: "Analyses of the accumulating study data are performed at prospectively planned timepoints within the study, can be performed in a fully blinded manner or in an unblinded manner, and can occur with or without formal statistical hypothesis testing."

This definition did not require full control of Type I error although it certainly does not imply that such control is unimportant. It did restrict itself to changes made "based on analysis of data … from subjects in the study," thus excluding changes made due to outside influences. An interesting example of changes based on outside influences could be a change to the primary endpoint (surely seen as a very major change). If this were done in an ongoing double-blind study because a completely separate study in the same indication had suggested an endpoint other than the intended primary endpoint were more appropriate, then many would argue there is no risk in changing the endpoint for the ongoing study. This would be in stark contrast to a study that had completed (or was ongoing but unblinded) and where the primary endpoint failed to show any treatment effect; or even in an ongoing and fully blinded study but where the primary endpoint event rate was very different to what had been expected at the planning stage. A subsequent change to a different endpoint, and then declaring that as "primary," would clearly risk introducing unacceptable bias.

The FDA definition also referred only to changes based on analyses "at prospectively planned timepoints." This is important and generally considered as the ideal conditions under which adaptations should be made. In particular, adaptive designs are seen as offering potential solutions to pre-thought-out potential problems, not as reactive fixes to problems that were unexpected.

Although the guidance was issued as a draft, in September 2018 it was updated (although still remained in draft form) to become "Adaptive Designs for Clinical Trials of Drugs and Biologics." Thereafter it was finalized in November 2019. In the latter two versions of the document, the definition was slightly changed, as follows:

For the purposes of this guidance, an adaptive design is defined as a clinical trial design that allows for prospectively planned modifications to one or more aspects of the design based on accumulating data from subjects in the trial.

Apart from cosmetic changes such as "study" becoming "clinical trial," perhaps the most striking difference is that reference to possible changes to the *study hypothesis* has been removed from the definition. While it seems unlikely that a blanket endorsement of changes to study hypotheses, midway through trials, is being advocated, more likely the authors are encompassing *study hypotheses* within the framework of *aspects of the design*. Importantly, it is stated that the definition of "interim analysis" is being taken far more widely than it is in the ICH E9 (Statistical Principles for Clinical Trials) document—and this has probably led to dropping the phrase "usually interim data" from the original definition.

With reference to FDA guidance, it is noteworthy that the Center for Devices and Radiological Health (CDRH) and the Center for Biologics Evaluation and Research (CBER) also jointly issued guidance in July 2016 on "Adaptive Designs for Medical Device Clinical Studies." Their definition is as follows:

"An adaptive design for a medical device clinical study is defined as a clinical study design that allows for prospectively planned modifications based on accumulating study data without undermining the study's integrity and validity."

And a footnote clarifies that:

"Integrity refers to the credibility of the results and validity refers to being able to make scientifically sound inferences."

One might conclude that these different FDA definitions are all similar but have slightly different emphasis and it is unclear—particularly with regard to words and phrases that are omitted—just how widely the agency is thinking on this topic.

Aside from the major regulatory agencies, another definition has been proposed which is probably the most widely quoted—this is from the Pharmaceutical Research and Manufacturers of America (PhRMA). They propose:

A clinical trial design that uses accumulating data to decide on how to modify aspects of the study as it continues, without undermining the validity and integrity of the trial. [1,2] This definition seems most broad. It makes no restriction about control of Type I error; it makes no restriction about preplanned versus post-hoc adaptions. It hints at data coming from within the study (accumulating data) but does not exclude the influence of data external to the study.

How does an adaptive design work?

It seems a very simple idea to watch and monitor the progress of a clinical trial and make changes to the design as it proceeds. In essence, this is the intent of the adaptive designed clinical trial, although the practicalities are far more complex. Indeed, it seems almost that we are turning our backs on well-learned lessons from the last 50 or more years since Bradford Hill carried out his classic experiment testing the treatment of tuberculosis with streptomycin.

There are many trials that—for many years—have incorporated some form of "adaption," although that term may never have been used to describe them. Sequential trials (or more commonly, group sequential trials) incorporate one or more interim analyses where a decision is made to continue or stop a trial. Statistical "stopping rules" have been widely developed to account for the problem of interpreting P-values when multiple significance tests on accumulating data have been carried out. Independent data monitoring committees then typically make recommendations to the sponsor on whether the trial should continue or not. These data monitoring committees are—contrary to the usual accepted rules of a double-blind trial—unblinded to the treatment allocation and see accumulating efficacy and safety results, although the sponsor does not. The decision to stop a trial early (if that is the recommendation that is made) makes the trial (by some definitions) adaptive: the intended course of the trial can change along the way. Whether a group sequential study design should be called adaptive depends on your view as to whether complete termination of a study fits within the realms of modifying "aspects of the study *as it continues*." The emphasis has been added but, clearly, a study stopped at an interim analysis is no longer continuing.

Introduced more recently has been the idea of "sample size re-estimation." Here, part way through recruitment to a trial, an updated estimate of how many patients will be needed (in total) to preserve some specified level of statistical power is made. The total sample size may not, therefore, be as was originally planned or expected: again, this is a form of adaption. This type of method is usually implemented without unblinding the trial and is usually quite uncontroversial—although many people consider it not really an adaptive design because the study has not been unblinded; in other situations it is done on unblinded data and, as with group sequential methods, independent unblinded data monitoring committees are usually employed

Such "seamless" designs need to be considered in one or other category: operationally seamless or inferentially seamless. Operationally seamless means that the whole protocol is set out in advance and successive patients can be recruited without any interruption as time goes by. It is assumed that drug supply (possibly different doses, availability of active comparator, etc.) are all available as soon as required and as soon as decisions have been made about how the trial will continue. Inferentially seamless implies that the inference(s), or conclusions from the study, will be based on the entire dataset; that is, on all patients who take part in the study. In effect, the Phase II patients are used twice: once for dose determination and again for confirmation of effect. In contrast to this (using the example illustrated in Fig. 4.1) we might use only the data from the part of the study akin to the "Phase III part" for making final conclusions. Such an approach might avoid some of the bias in estimating the size of treatment effect described earlier in this section. A study can be operationally seamless with or without being inferentially seamless. If we do not try to make a trial inferentially seamless, then—although it is one ongoing trial—it bears much closer analogies to a Phase II trial followed by a Phase III trial and maintains the ideas of distinguishing between a learning phase and a confirming phase.

Platform trials

A further development from the so-called "seamless" designs is that which comes under the broad description of "platform trials" or "master protocols." The idea is that several related trials can all be conducted under one protocol. Despite added complexity in writing such protocols, it can bring efficiencies in terms of protocol approval processes, ethics approval, regulatory approval, and so on. In addition, there can be a scientific advantage because a trial is not necessarily restricted to two (or a few) predetermined treatments, but can open out to include new therapies as they become available, and drop existing (trial) therapies that show little promise. Yet, during such a process, the one trial keeps going under the same protocol. So it should be clear from this that the focus of a platform trial is really about studying a disease (or condition) rather than studying a specific treatment. A further (although rather different) purpose could be to keep the treatment(s) fixed but change the patient population to try to determine which types of patients (perhaps driven by a biomarker) might be most suited to a particular treatment. All these options essentially boil down to learning something new within a trial, then making changes to that trial based on what we have learnt: essentially following the general definitions of adaptive trials given earlier.

It is worth distinguishing two distinct types of platform trials. The terminology is perhaps arbitrary but often gets confused. A clear distinction is made in Ref. [5], from where we summarize, as follows:

• Basket trials or bucket trials investigate multiple diseases based on a common molecular target in a single trial usually with one investigational drug. All

screened patients expressing the same molecular target are enrolled in the trial and are treated with the same investigational targeted drug therapy.
• Umbrella trials are generally disease specific and investigate different molecular targets for one disease in parallel substudies applying biomarker stratification. Substudies are usually randomized and compare each target therapy against standard of care or placebo. Patients expressing none of the screened targets may be stratified to a biomarker-negative control cohort.

Two notable examples of platform trials are the I-SPY collection of trials (https://www.ispytrials.org) and the STAMPEDE trials (http://www.stampedetrial.org). Many peer-reviewed publications have stemmed from these projects (particularly so for I-SPY) but their respective websites give far fuller and up-to-date information about them. A publication from the Adaptive Platform Trials Coalition [6] lists other trials and gives a brief summary of some of their defining features, illustrating the great variety of design aspects that might be considered: features such as phase, duration, number of treatments, relative assignment to different treatment arms, timing of endpoints, interim analyses and frequency of interim analyses, statistical approaches, and so on.

Objections to adaptive designs

The CHMP Reflection Paper makes a particular point that "the purpose of phase III is to confirm the findings from preclinical studies, tolerability studies, dose-finding and other phase II studies … To argue for design modifications in a phase III trial (or a late stage phase II trial supposed to be part of the confirmatory package) is then a contradiction to the confirmatory nature of such studies." Indeed, the document is titled "Methodological Issues in *Confirmatory* Clinical Trials Planned with an Adaptive Design" [emphasis added]. It has been well-recognized that post-hoc analyses of trial results can lead to substantial bias in their interpretation. Hence, there is very strong emphasis made in, for example, ICH E9 about prespecifying analyses and Statistical Analysis Plans. Now, apparently, we have a trial design that allows us to change design, endpoint, timing of assessment, choice of comparator, method of analysis (the list is seemingly endless) and yet still regard this as a confirmatory trial rather than an exploratory one. How can this be?

Perhaps the simple answer is that adaptive designs are not the panacea for treatments that do not work, or for poorly designed trials or development plans. Unfortunately, in the earlier days of using these sorts of trials, many proposals were methodologically very poor and paid little attention to things like operational bias in ongoing studies or correct analyses (either significance testing or estimation). However, even today, operational bias is often ignored—perhaps because it is difficult to quantify. Most of the research work on adaptive designs has been carried out by statisticians. As a consequence, the issues of correct significance testing and estimation have taken prominence. But as outlined at the beginning of the chapter, there is an element of ignoring long-learnt lessons from more than 50 years' experience of conducting clinical trials. To help circumvent this, data monitoring committees are being used more and more to inspect interim results and make recommendations for changes. Regulators (and sponsors) have a lot of experience with operational aspects of data monitoring committees—and regulatory guidance is extensive. However, it has to be realized that data monitoring committees are being asked to make much bigger decisions in adaptive designs than in more traditional interim analyses: they may be recommending changes to inclusion criteria, endpoints, choice of comparator, or others. In a commercial setting, all of these points can have major implications on an eventual product label, and such decisions cannot be made independently of company business strategy. Yet as soon as the sponsor is involved in making these decisions, the study may start to become unblinded. There is a potential paradox that the "more adaptable" a trial can be, the less acceptable it may be to the sponsor who cannot take part in the decision-making.

Potential gains from adaptive designs

Objections and difficulties should not be dismissed easily, but they should be managed so that the potential gains from such designs can be realized. The obvious gain is that a trial of a useful medicine, but that is not optimally designed so as to demonstrate the benefits of that medicine can be changed along the way. This avoids the possibilities of (at best) having to redo trials with the ensuing costs and delays in getting good medicines to patients. At worst it avoids the possibility that a good medicine might be dismissed as ineffectual and discarded when, in fact, it was the trial that was lacking, not the medicine.

The further time-saving opportunity is illustrated in Fig. 4.2. Part (A) of Fig. 4.2 illustrates a typical (if simplified) example Gantt chart for a drug development program. It begins with Phase I (perhaps several studies; other clinical pharmacology studies also might be carried out later during the development period). We then have Phase II (here shown as just one study) and then Phase III (shown as two studies, as is often the case). The important point to note is not the time when studies are ongoing, but the so-called "white space" between the end of each phase and the beginning of the next. Much time is used here with administrative issues such as regulatory and ethics approvals to carry out the next trial, as well as study "start-up" activities such as training investigators, preparing courier services, agreeing contracts, and many more. It is these "white spaces" that are a great opportunity to eliminate. Part (B) of Fig. 4.2 shows the same duration of trials, but Phases II and III are put together, back-to-back, illustrating how (operationally) Phase III will run seamlessly off the back of Phase II. However, the challenge and time taken to write the protocol for some of these adaptive (and platform) trials should not be underestimated. It is almost inevitable that it will take longer to *start* an adaptive designed trial than to start a more traditional designed trial; the challenge is to be able to make up that delay and reach an ultimate conclusion more quickly.

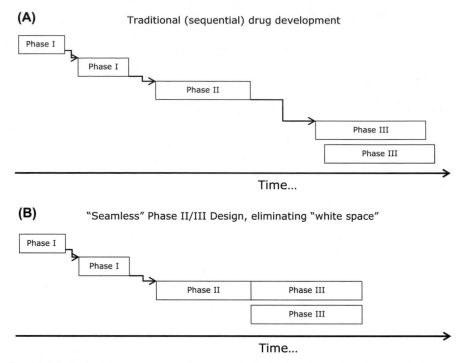

(A) Traditional (sequential) drug development

Phase I
Phase I
Phase II
Phase III
Phase III

Time...

(B) "Seamless" Phase II/III Design, eliminating "white space"

Phase I
Phase I
Phase II   Phase III
Phase III

Time...

FIGURE 4.2 Typical Gantt chart showing a sequence of trials in the drug development plan. Part, (A) Traditional (sequential) drug development, Part (B) "Seamless" Phase II/III Design, eliminating "white space."

The transition between the different Phase I trials, and from Phase I to Phase II may still follow a very traditional sequential path, with planning time (or "white space") between them (as illustrated in both scenarios). Scenario (B) shows the white space eliminated between Phase II and Phase III—an opportunity provided via the use of a seamless Phase II/III adaptive design—and the resulting time saved until completion of the Phase III (and full development) program.

SWOT analysis

Strengths
➢ Knowledge of what is happening in a trial should be far more reliable than prior assumptions.
➢ Trials that are potentially failing (as well as treatments that are potentially failing) can be identified much sooner than from a conventional fixed design.

Weaknesses
➢ More time must be invested when designing a trial to map out potential adaptions that might be needed.
➢ Not all bias—particularly operational bias—can necessarily be eliminated from an adaptively run trial.

SWOT analysis—*cont'd*

Opportunities
➢ Fewer trials should fail to answer the question they were designed to answer.
➢ Effective therapies, tested in ineffective trials, should be a thing of the past if we can eliminate ineffective trials.
➢ Software for simulating and implementing adaptive designs have been very quick to develop, allowing much easier design of adaptive trials.

Threats
➢ There have been many cases of poorly designed or poorly implemented adaptive trials.
➢ Bad examples have made regulators very cautious, leading to a very conservative approach from many sponsors.

Applicable regulations

ICH E3 (Structure and Content of Clinical Study Reports); ICH E6 (Guideline for Good Clinical Practice); ICH E8 (General Considerations for Clinical Trials); ICH E9 (Statistical Principles for Clinical Trials); ICH E10 (Choice of Control Group and Related Issues in Clinical Trials).

FDA Guidance for Industry "Adaptive Design Clinical Trials for Drugs and Biologics" (draft, February 2010), Adaptive Designs for Clinical Trials of Drugs and Biologics (draft, September 2018, and final, November 2019); FDA Guidance for Industry "Adaptive Design Medical Device Clinical Studies" (July 2016); FDA Guidance for Clinical Trial Sponsors "Establishment and Operation of Clinical Trial Data Monitoring Committees" (March 2006); "Master Protocols: Efficient Clinical Trial Design Strategies to Expedite Development of Oncology Drugs and Biologics" (draft, September 2018).

CHMP Reflection Paper on Methodological Issues in Confirmatory Clinical Trials Planned with an Adaptive Design (CHMP/EWP/2459/02, October 2007); CHMP Guideline on Data Monitoring Committees (EMEA/CHMP/EWP/5872/03 Corr, July 2005).

Take-home message

➢ The term "adaptive design" is very broad.
➢ Adaptive designed trials allow the possibility to monitor the progress of an ongoing trial (in a blinded or unblinded manner) and make appropriate design changes as the trial progresses.
➢ Such possible design changes should generally be prespecified in the protocol and not simply reactive fixes (what would be an amendment).
➢ Wide-scoping trials comparing many treatments for a single medical condition (umbrella trials) are also more commonly appearing.
➢ There are many technical, statistical, as well as operational challenges to planning and running such studies. Identifying and solving problems of bias are particularly paramount and data monitoring committees are being used more and more to try to help with this problem.

References

[1] Gallo P, Chuang-Stein C, Dragalin V, Gaydos B, Krams M, Pinheiro J. Adaptive design in clinical drug development — an executive summary of the PhRMA Working Group (with discussions). J Biopharm Stat 2006;16:275—83.

[2] Gallo P, Krams M. PhRMA Working Group on adaptive designs: introduction to the full white paper. Drug Inf J 2006;40:421—3.

[3] O'Quigley J, Pepe M, Fisher L. Continual reassessment method: a practical design for phase 1 clinical trials in cancer. Biometrics 1990;46:33—48.

[4] Bauer P, Köhne K. Evaluation of experiments with adaptive interim analyses. Biometrics 1994;50:1029—41.

[5] Sudhop T, Brun NC, Riedel C, Rosso A, Broich K, Senderovitz T. Master protocols in clinical trials: a universal Swiss army knife? Lancet Oncol 2019;20:e336—342.

[6] The Adaptive Platform Trials Coalition. Adaptive platform trials: definition, design, conduct and reporting considerations. Nat Rev Drug Discov 2019;18:797—807.

A vision: in silico clinical trials without patients

Lesley A. Ogilvie[1,2], Damian T. Rieke[3,4], Hans Lehrach[1,2,5,6]

[1]*Alacris Theranostics GmbH, Berlin, Germany;* [2]*Max Planck Institute for Molecular Genetics, Berlin, Germany;* [3]*Charité Comprehensive Cancer Centre, Charité — Universitätsmedizin Berlin, Berlin, Germany;* [4]*Berlin Institute of Health, Berlin, Germany;* [5]*Dahlem Centre for Genome Research and Medical Systems Biology, Berlin, Germany;* [6]*Southern University of Science and Technology, Shenzhen, Guangdong, China*

The need

Many of the problems faced in healthcare are based on a simple fact: we are all biologically different and so are our diseases. It is therefore not surprising that drugs often only help a fraction of the patients receiving them and patients can have (sometimes life threatening) adverse reactions to the drugs they take. In Europe alone, close to 200,000 patients die every year due to adverse drug reactions [1,2], a major contributing factor to the ever-increasing costs of healthcare (4.5 billion euros every day in Europe) [3], and an obvious problem in light of ongoing demographic change [4].

The biological differences between patients and their diseases with the same pathology are, however, also a major factor underlying the difficulties we have in the development of new drugs, with development times, risks, and costs per approved drug rising to an increasingly unsustainable level [5]. In addition, many approved drugs only help a fraction of the patients receiving them, often resulting in (on average) moderate increases in survival times [6,7], despite costing in excess of $100,000 (€85 000) per year of treatment [8].

The ultimate test of the clinical usefulness of a drug currently takes place in the last stage of drug development, at a point when massive resources (money and time) have already been invested. If a drug fails in a phase III clinical trial, the financial impact can be severe, even threatening the survival of large pharmaceutical companies. A major reduction of both the risk and cost associated with the drug development process could be achieved through conducting in silico clinical trials at the beginning of the drug development process to eliminate drug candidates most likely to fail in the late stages of development. Virtualization (in silico) can, however, also significantly improve and accelerate the other phases of drug development. Potentially, the majority of preclinical work could be carried out on computer models, with real experiments used to validate the results generated from modeling. This will not only accelerate the preclinical development phase but also reduce costs

Innovation in Clinical Trial Methodologies. https://doi.org/10.1016/B978-0-12-824490-6.00020-7

and the number of experimental animals needed. It will also help significantly to increase the accuracy of the extrapolation from the experimental system, since the results with the preclinical system can be automatically translated to predict the response of the human model and therefore of the human patients. In silico trial results could also help guide replacement of large and expensive (and often inconclusive) real-life clinical trials with small, quick, and relatively cheap real trials on patients, selected as responders (on the basis of in silico trial results).

In our view, a major part of the problem is the assumption we usually start with: that patients with a specific disease are very similar, forming a homogeneous population. This assumption underlies the concepts of first-line treatment and the blockbuster drug in pharmaceutical development, both of which, however, have proved inadequate to explain the response of real patients to real drugs. In a process reminiscent of the Ptolemaic world view (and his introduction of an increasing number of epicycles to explain the increasing differences between concept and observation), this has led to increased stratification of patient populations into subgroups, which in turn are assumed to be homogeneous (but typically are not).

We argue here for a completely different ("Copernican?") view: ultimately all patients and all diseases are likely to be different.

The solution

The problem of complexity is one we face in many areas of our lives, not just healthcare. In complex situations, mistakes are inevitable; however, in all areas except medicine and drug development (and other biological problems), we have learned to avoid their consequences by making them safely, cheaply, and quickly on computer models of the real situation rather than in reality. We do not conduct crash tests with real cars in the initial stages of safety testing; instead, large numbers of virtual crash tests are conducted first. We do not train pilots on planes full of passengers, but let them make their mistakes during training in flight simulators. Similarly, we should first test therapy options on digital twins of real patients to identify the individually optimal therapy or therapy combination, and test drugs first in in silico clinical trials, carried out on digital twins of real patients, to identify subgroups of patients most likely to respond to the drug. This can be done before entering a drug candidate into preclinical development, followed by small, real-life clinical trials on patients identified as responders by their digital twins [9].

This raises the obvious question of how real clinical trials can be performed when all attempts at stratification still leave us with very heterogeneous patient groups. The least heterogeneous group is the single patient, raising serious problems for the general concept of evidence-based medicine. The obvious solution for this would be to abstract the concept of evidence-based medicine further, moving from clinical trials testing the response of (necessarily heterogeneous) groups of patients to a specific drug, to testing the response of the same group of patients to whatever drug or drug combination their digital twins respond to best (Fig. 5.1), thus including the testing

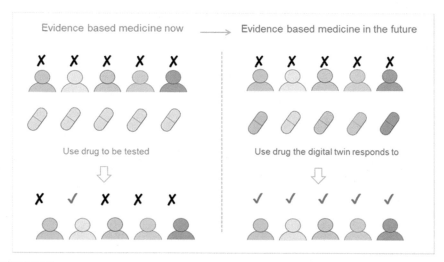

FIGURE 5.1

Conducting in silico clinical trials: evidence-based medicine of the future. Drugs tested within clinical trials will be selected based on the outcomes of modeling and which drug or drug combination the digital twin responds to best. Inclusion of the testing and treatment allocation process into clinical trials will allow for evidence-based comparisons between treatment strategies (e.g., a phase III trial randomizing the left vs. right panel).

and treatment allocation process into the clinical trial. In doing so, these novel drug allocation strategies can be tested against conservative treatment approaches and prove their benefit within the concept of evidence-based medicine.

The obvious next question is: How can we construct such digital twins of individual patients? As example, in oncology, we need in silico models of tumors that take into account all information on tumor heterogeneity, as well as enough of the biology of the patient (e.g., genome, relevant metagenomes, the status of the immune system, etc.) to be able to classify variants as somatic or germ line, predict how a drug is metabolized in the intestine and liver, identify potential side effects of the drug in normal tissues, and, if relevant, predict the response of the immune system to the tumor. For diseases like cancer, dominated by molecular and cellular processes, these models would be in the form of interacting molecular models, with relevant drugs modeled as molecules which interact with their molecular targets in the relevant cells or tissues, e.g., through affecting the concentration of the active form(s) of the drug. For disease areas with strong physiological (e.g., cardiovascular diseases) or neuronal causality (e.g., neurological diseases), these relevant mechanisms would obviously have to be incorporated as well.

To establish such molecular models, we need three components:

1. The structure of the relevant biological processes (e.g., the signaling pathways in the tumors controlling cell division and cell death), studied extensively by basic research in this area, represented in an object-oriented reference model;

2. Extensive molecular characterization of the individual tumor and patient, made possible through the enormous progress in technology, e.g., next generation sequencing and other molecular analysis techniques, used to generate a personalized model representing the individual tumor and patient;

3. The personalized model, including new objects representing the drug or drug combination, translated into large systems of differential equations, which can be solved numerically if (and only if) we have quantitative values for the parameters representing the (mostly unknown) rate and equilibrium constants, as well as unknown starting values for specific components.

While we have a pretty good idea of the model structure required for the first step (see Refs. [10−14]), and are increasingly able to determine the individual molecular data of tumors and patients required for the second step [15], we have not had the ability to determine the missing parameter values that are critical for deriving quantitative predictions from the personalized models. Recently, however, we have had very encouraging results for this last step, potentially representing the last major bottleneck on the way to a data- and model-driven truly personalized medicine, and a largely virtualized drug development process.

For this, we use detailed reference models of cancer-related signaling networks using PyBioS, a web-based platform for modeling of complex molecular systems [10−14]. Our current models represent relevant biological processes through an object-orientated mechanistic model of cellular signaling, which can be personalized by omics data from a patient and their tumor. Drugs are also represented as objects, interacting with their target object, e.g., to form an inactive complex, simulating the response of the tumor to drug treatment.

Large cohorts of these computer models ("digital twins") can be assembled for testing the effectiveness (and potentially safety) of drugs (singly and in combination) at all stages of the drug development process (both preclinical and clinical), with associated benefits, ethically, clinically, and financially (Fig. 5.2).

SWOT analysis

Strengths

Through assembly of large cohorts of computer models, a much more comprehensive representation of the biological diversity of patients and their potential contraindications could be gained to test the effectiveness of drugs (singly and in combination), much more quickly, safely, and cheaply. For instance, publicly available omics data from large-scale cancer sequencing projects, such as The Cancer Genome Atlas (TCGA) and the International Cancer Genome Consortium (ICGC), provide a solid data foundation for creating individualized models, with further data accumulating as sequencing costs steadily decrease. In silico cohorts consisting of thousands of individual models could be used to simulate multiple drug combination scenarios within days, as compared to months (or even years)

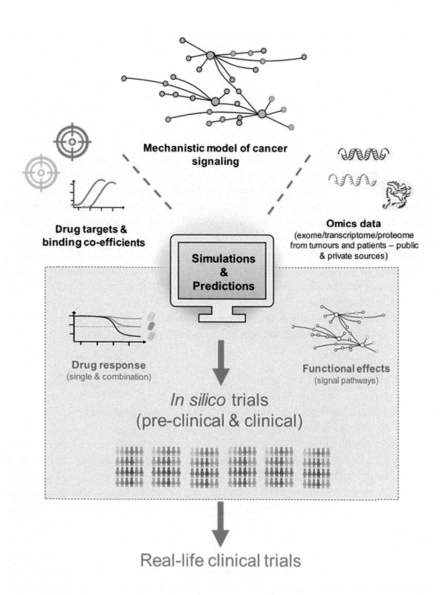

FIGURE 5.2

Digital twin technology for pre/clinical oncology trials.

for real-life clinical trials. The results generated would provide crucial evidence and rationale for choosing specific combinations that are more likely to be successful for an individual, based on their molecular profiles. Given the increasing importance of combination regimes for complex diseases such as cancer (combination treatments account for $> 25\%$ of clinical trials in oncology) [16], the ability to rapidly simulate effects on large cohorts and demonstrate whether the combination is safe and effective represents a major strength of this approach.

Weaknesses

Although our current generic mechanistic model [11–14] uniquely reflects a complex single cell signaling transduction network, it does not address all key features that might impact drug resistance such as tumor heterogeneity, cell-cell interactions, and the impact of the tumor microenvironment and immune pathways; however, model development is ongoing to incorporate such elements. Current work is focused on the development of an integrated tumor-patient model that includes factors that impact an individual's response to drugs, such as the immune system, pharmacogenomics, the gut microbiome, and the ability to identify potentially unacceptable side effects, by including other cell types and tissues.

In situations where there is no information available on exact molecular mechanisms, mechanistic model components could be combined with Artificial Intelligence (AI) techniques to generate "hybrid" models. To incorporate nonmechanistic components into our mechanistic models, we can define "pseudo-objects" as part of the object-oriented modeling system encapsulating, e.g., neuronal or Bayesian networks trained to translate the results of the mechanistic model into complex consequences by processes for which no molecular mechanisms have been identified.

Opportunities

Virtualization of the drug development process, including in silico clinical trials, can potentially lower the overall costs of developing drugs by orders of magnitude, through improved targeting of small, well-defined patient groups, acceleration of the process, reduced dropout rates, and replacement of a large fraction of expensive preclinical and clinical work by quicker and cheaper in silico experiments, leading both to lower costs for the healthcare system and a significant increase in the number of new drugs reaching the market.

In silico trials have the potential to be an integral part of the whole drug development process, both within the preclinical and clinical phases. For example, as soon as a drug's binding affinity to a molecular target can be accurately predicted, models can be used to make predictions and focus on the next stage of development. Alignment with pharmaceutical early stage development would provide scope to identify potential responder groups (through identification of biomarkers) using patient data (both public or from industry), guiding preclinical experimental validation efforts.

Rescuing and repurposing drugs that have failed in conventional trials due to low response rates rather than toxic side effects becomes a systematic process, as does the identification of effective drug combinations offering high efficacy and reduced adverse effects through in silico screening of large numbers of drugs (including their additive and synergistic effects). Replacement of preclinical phases, in total or in part, by in silico trials also opens up opportunities for a significant reduction in animal testing, providing an ethically rigorous option that saves costs and time (Fig. 5.3).

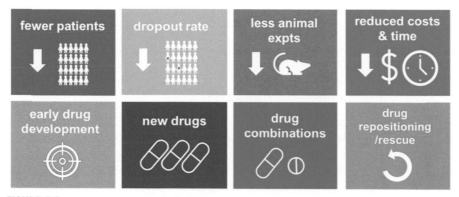

FIGURE 5.3

Opportunities and applications of in silico trials.

As we navigate this current transition period, in which technology development and affordability are starting to align, the prospect of routine genomic/molecular sequencing for clinical practice is being recognized as feasible and acceptable by healthcare systems and professionals, and the public in general. Initiatives such as the 100,000 Genomes Project in the UK [17] are demonstrating the benefits whole genome sequencing can have with regards to finding treatment options for patients with rare diseases and cancer. As molecular analysis becomes a routine part of our healthcare, so will the application of modeling approaches to translate these data into actionable insights and guide treatment (and prevention) options for individuals. Building on this, avenues of opportunity arise for conducting so called "n-of-1" clinical trials [18], in which the effectiveness of a drug could be assessed on an individual's specific molecular (and other) characteristics and the results of modeling. Indeed, new clinical trial designs that incorporate the developments in personalized medicine are already starting to be implemented, e.g., Basket and Umbrella trial designs for cancer, which use molecular features to assemble cohorts [19]. Accordingly, the first histology-agnostic cancer drugs have already been approved in Europe and the United States [20]. These developments form a foundation for the in silico trial approach, which could eventually replace costly phase III trials. The models themselves (of both patients and preclinical systems) would form an invaluable resource, evolving continually by learning from the differences between predictions and actual results, enabling improvement of the model structure and accuracy of predictions and providing insight into the mechanisms of response.

Threats
Ultimately, every new technology will have to prove its value. In silico modeling generates information which is subsequently rigorously validated experimentally and/or clinically. Additionally, a number of barriers exist that require credible

solutions to pave the way toward implementation. Barriers to adoption of the in silico trial concept include:

1. Resistance from users (from healthcare professionals to the general public)
2. Regulatory approval
3. Validation of predictive capacity
4. Support from pharmaceutical companies

Breaking down these barriers has been the aim of the Avicenna Alliance (https://avicenna-alliance.com), which was commissioned by the European Union to create a research and technological development roadmap outlining a strategy for in silico clinical trials, involving all stakeholders, from patients to clinicians to the pharmaceutical industry [21]. This includes a challenge to cut the cost of bringing a new pharmaceutical to market from 2.5 billion to 250 million US dollars by 2025 [22]. In recent years, there has also been a move toward acceptance of the concept by regulatory agencies. For example, the US Food and Drug Administration (FDA) is actively supporting the development of computer models for personalized medicine with the establishment (in 2017) of an FDA-wide working group on "Modeling and Simulation" focused on the integration of computer modeling into the FDA's review processes.

Our current work is also addressing key concerns by validating the predictive accuracy of the mechanistic models developed within the framework of national and international research projects and clinical studies (e.g., BMBF Treat20plus, ERA-PERMED PECAN, H2020 IPC & CanPathPro).

Take-home message

The development of the technology required to generate (molecular) digital twins provides the technological basis for in silico clinical trials, as well as major virtualization of many other steps in the drug development process, offering the potential of orders of magnitude reductions in the cost of bringing new drugs to the market, and a corresponding increase in new drugs in many disease areas, including those for which risks have been considered too high to justify commercial development. The same basic technology will, however, also be used to truly personalize therapy choice and preventive and wellness measures, further increasing health and well-being worldwide.

Acknowledgments

The authors would like to thank colleagues at Alacris Theranostics and the Max Planck Institute for Molecular Genetics, especially Dr. Marie-Laure Yaspo and Dr. Bodo Lange, for many discussions that have helped to form the concepts presented here. This work was supported in part by funding from the European Union's Horizon 2020 research and innovation program under grant agreement No. 826121 (IPC).

References

[1] Lazarou J, Pomeranz BH, Corey PN. Incidence of adverse drug reactions in hospitalized patients: a meta-analysis of prospective studies. J Am Med Assoc 1998;279:1200–5. https://doi.org/10.1001/jama.279.15.1200.

[2] Bouvy JC, De Bruin ML, Koopmanschap MA. Epidemiology of adverse drug reactions in Europe: a review of recent observational studies. Drug Saf 2015;38(5):437–53. https://doi.org/10.1007/s40264-015-0281-0.

[3] Eurostat, healthcare expenditure. Available from: http://ec.europa.eu/eurostat/web/health/health-care/data/database/.

[4] 2018 ageing report: policy challenges for ageing societies. Available from: https://ec.europa.eu/info/news/economy-finance/policy-implications-ageing-examined-new-report-2018-may-25_en. [Accessed 1 December 2019].

[5] The truly staggering cost of inventing new drugs. Matthew Harper; 2012. Available from: https://www.forbes.com/sites/matthewherper/2012/02/10/the-truly-staggering-cost-of-inventing-new-drugs/#597f10e34a94. [Accessed 1 December 2019].

[6] Davis C, Naci H, Gurpinar E, et al. Availability of evidence on overall survival and quality of life benefits of cancer drugs approved by the European Medicines Agency: retrospective cohort study of drug approvals from 2009–2013. BMJ 2017;359:j4530.

[7] Prassad P. Do cancer drugs improve survival or quality of life? BMJ 2017;359:j4528. https://doi.org/10.1136/bmj.j4528.

[8] Kantarjian H, Rajkumar SV. Why are cancer drugs so expensive in the United States, and what are the solutions? Mayo Clin Proc 2015;90:500–4. https://doi.org/10.1016/j.mayocp.2015.01.014.

[9] Lehrach H. Virtual clinical trials, an essential step in increasing the effectiveness of the drug development process. Public Health Genomics 2015;18(6):366–71. https://doi.org/10.1159/000441553.

[10] Wierling C, Herwig R, Lehrach H. Resources, standards and tools for systems biology. Brief Funct Genomic Proteomic 2007;6:240–51.

[11] Wierling C, Kühn A, Hache H, Daskalaki A, Maschke-Dutz E, Peycheva S, et al. Prediction in the face of uncertainty: a Monte Carlo-based approach for systems biology of cancer treatment. Mutat Res 2012;746(2):163–70. https://doi.org/10.1016/j.mrgentox.2012.01.005.

[12] Wierling C, Kessler T, Ogilvie LA, Lange BM, Yaspo ML, Lehrach H. Network and systems biology: essential steps in virtualizing drug discovery and development. Drug Discov Today Technol 2015;15:33–40. https://doi.org/10.1016/j.ddtec.2015.07.002.

[13] Röhr C, Kerick M, Fischer A, Kühn A, Kashofer K, Timmermann B, et al. High-throughput miRNA and mRNA sequencing of paired colorectal normal, tumor and metastasis tissues and bioinformatic modeling of miRNA-1 therapeutic applications. PLoS One 2013;8(7):e67461. https://doi.org/10.1371/journal.pone.0067461.

[14] Fröhlich F, Kessler T, Weindl D, Shadrin A, Schmiester L, Hache H, et al. Efficient parameter estimation enables the prediction of drug response using a mechanistic pan-cancer pathway model. Cell Syst 2018;7(6). https://doi.org/10.1016/j.cels.2018.10.013. 567–579.e566.

[15] Schütte M, Ogilvie LA, Rieke DT, Lange BMH, Yaspo ML, Lehrach H. Cancer precision medicine: why more is more and DNA is not enough. Public Health Genomics 2017;20(2):70–80. https://doi.org/10.1159/000477157.

[16] Wu M, Sirota M, Butte AJ, Chen B. Characteristics of drug combination therapy in oncology by analyzing clinical trial data on ClinicalTrials.gov. Pac Symp Biocomput 2015:68—79.

[17] The 100,000 genomes project. Available from: https://www.genomicsengland.co.uk/about-genomics-england/the-100000-genomes-project/.

[18] Schork NJ. Personalized medicine: time for one-person trials. Nature 2015;520(7549): 609—11. https://doi.org/10.1038/520609a.

[19] Janiaud P, Serghiou S, Ioannidis JPA. New clinical trial designs in the era of precision medicine: an overview of definitions, strengths, weaknesses, and current use in oncology. Cancer Treat Rev 2018;73. https://doi.org/10.1016/j.ctrv.2018.12.003.

[20] https://www.fda.gov/drugs/resources-information-approved-drugs/fda-grants-accelerated-approval-pembrolizumab-first-tissuesite-agnostic-indication.

[21] Avicenna alliance. Available from: https://avicenna-alliance.com/about-us/avicenna-roadmap/.

[22] Viceconti M, Kennedy J, Henney A, Reiterer M, Polak S, Colaert D, et al. (Avicenna Consortium) in silico clinical trials: how computer simulation will transform the biomedical industry. 2016. https://doi.org/10.13140/RG.2.1.2756.6164.

Studies with less patient interaction

The patient as sub-investigator

Brendan M. Buckley, MD, DPhil, FRCPI
Chief Medical Officer, Teckro, Limerick, Ireland

The need

People that participate in clinical trials as subjects often do so at significant inconvenience, particularly if the trial is of a preventive intervention or is in stable chronic disease. Simple logistical barriers exist which they have to overcome, such as taking time off work, traveling to the trial site, and dealing with the nontrivial matter of finding car parking when they get there. A trial may sometimes extend over several years; so it is not surprising that dropout rates may be large, sometimes to the extent that the trial is compromised. Regrettably, the subject is often regarded by study teams as literally that, somebody subject to the requirements and edicts of the protocol and of those who implement it. Lip service may be given to the idea that the subject is a true stakeholder, but this is frequently not really acted on. While it is nowadays fashionable to speak of "patient-centricity," the general public still describes trial subjects as "guinea pigs," uncomfortably often. Are the true attitudes of those commissioning and conducting trials all that far behind?

It is clear that many protocols contain requirements for the collection of large amounts of information that never contribute to results or conclusions. There are many consistent anecdotes about this, but a study that evaluated the case record forms (CRFs) in eight oncology trials and their nine resulting publications is reflective [1]. The CRF analysis revealed that the total collected data items per subject ranged from 186 to 1035 per trial with a median of 599. Across all of the nine publications resulting from the trials, a median of 96 data items (18%) were reported in each manuscript, ranging from 11% to 27% per trial. When data items were subclassified, only 4% or less of collected data items were used in 8 of the 18 categories. This is not alone spectacularly wasteful of time and resources but is also an imposition on subjects, who might have misgivings about consent were they to realize it was going to happen. This waste is compounded by the mass effort of monitoring and correction by clinical research associates (CRAs) of trial data collected that are never used or reported.

Innovation in Clinical Trial Methodologies. https://doi.org/10.1016/B978-0-12-824490-6.00005-0

The solution

A key to minimizing much of this wasted effort is to ensure that the maximum proportion of observations collected enter the study database as primary source data. This will tend to cut out collection of less relevant, unnecessary data. It also should minimize the need for monitoring since these source data don't need reconciliation with anything else. The best way to accomplish this is to get the trial subjects to collect and report their own data directly, with no intermediary. In effect, this requires the subject to be given a sub-investigator role.

Many, if not most, patients with chronic stable disease are capable of generating and reporting their own data and, if the technology is appropriately designed, can do so with the same level of fidelity as can staff at a trial site. In this, the subject is acting as a true sub-investigator who is thereby much more deeply invested in the study and should be able to participate at much less inconvenience. I specify chronic stable disease in particular, as it is clear that studies of acute or unstable disease or studies that require complex clinical assessment or imaging are still required to be done at trial sites. By using appropriate connected health technologies that fully satisfy data standards such as 21 CFR Part 11 and the guidelines on electronic source data [2—4], subjects may collect and report their own data, including clinical measurements and ePRO-type diaries, in a manner that ensures that a large part of the data collected in the trial does not require to be monitored since it is primary "Source."

A typical opportunity is provided by a cardiovascular (CVS) outcomes trial with a new diabetes drug. Such trials are, in effect, mandatory unless sufficient data can be accumulated from elsewhere in the clinical development program to satisfy the requirement to show CVS safety. This kind of trial has a simple objective—to test the hypothesis that subjects randomized to the new drug have no more CVS adverse events than do subjects randomized to placebo against the background of best conventional standard of care. Conventionally, the designs of such trials are near replicas of all the cardiovascular outcome studies that have changed very little over several decades. They require tens of thousands of subjects in a massive exercise. Typically, they demand that subjects attend trial visits every few months, where blood for glucose, HbA1c, and lipid measurements are taken, as well as a variety of other parameters of largely academic interest ("It would be nice to measure …"). Trial visits also collect data on BP and ECG, as well as items such as lists of concomitant medications. Ultimately, however, all that matters in testing the hypothesis of the trial is to determine at the end whether there was an imbalance in CVS events between the two groups. Secondary outcomes, such as whether the drug helped achievement of diabetic control, are also important in a broader sense.

Diabetes cardiovascular outcome trials could easily be designed that are based on intensive collection of observations on themselves by the diabetic trial subjects. Sites would recruit subjects who would each be explicitly given the formal role of capturing key data on themselves, effectively making them sub-investigators. During the trial, they would measure their own blood glucose and BP several times a day and transmit the results from their measuring devices, for example, via an intermediary Bluetooth-linked smartphone to a central monitoring database. Other data could be collected, such as ePRO and measurements from the "internet of things," such as wireless temperature sensors in their refrigerator, if relevant. These data would be received by the clinical staff at their site and transformed by tracking algorithms to allow subjects' progress to be followed. These data, having been anonymized, would be accessible by centralized data monitoring staff from the sponsor or clinical research organization (CRO). As e-source data, they would be included in their eCRFs. In addition, the subjects/sub-investigators would have access to enable them to visualize their own progress. Those subjects submitting observations indicating suboptimal disease parameters, for instance, unsatisfactory diabetic control or blood pressure, would be communicated with directly by their trial site for assistance, fulfilling the requirements of ICH-GCP 4.3 and the Declaration of Helsinki. When clinical outcomes are as clear as are most of those in CVS trials, patients can easily report most hard trial endpoints themselves. In-person visits to the trial sites are reduced to those in which physical examination or other direct contact procedures are unavoidable. Capillary blood for essential tests that the subjects can't do themselves can be sent on filter paper to a central laboratory. ECGs may be done elsewhere than at the trial site if there is a more convenient location for the subject. It can be estimated that only about 15% of subjects would need to be called to visit the trial sites after trial entry to fulfill the requirements for good patient care. As the other 85% are constantly being monitored remotely, their care should be no less, and the investigator's clinical care responsibilities would be fulfilled.

The advantages of this approach are obvious. The density of the data on blood glucose would allow conclusions to be drawn regarding diurnal patterns of control and potentially allow each subject to be followed according to their own normative values rather than just by reference to the population distribution. The requirement for on-site source data monitoring would fall substantially as most data would be "source." Subject withdrawal is likely to be lessened due to improved convenience and better connection with the study. The net result is likely to be considerably less cost and greater speed of execution of the trial. It is easy to estimate a saving of 60%–70% on what is normally an extremely costly exercise when done conventionally.

No new technology needs to be invented to allow execution of a connected health trial; everything already exists on the market. However, there needs to be further developed a new and different role in sponsor and CRO organizations to allow centralized monitoring of large data streams on individual subjects. This role is being developed to an extent to facilitate risk-based monitoring, but it needs further work and change management. All those involved in sponsoring, designing, and executing trials will need to recognize that the classic way of doing business in chronic disease trials is wasteful and needs radical change. Covid-19 has sent a profound shock to the concept that it is OK to continue to operate a decades-old model whereby site-centric trials are seen as the only safe way to conduct business. The world changed and we were caught out talking about but still resisting change.

SWOT analysis

Strengths: Source data directly from patients will change traditional ways in which trials are monitored and decrease its necessity. Data will be richer, denser, and in considerably greater volume and will contribute to big data analytics systems.

Weaknesses: There are challenges for standardization and regulation of devices, analytical algorithms and systems in clinical trials in a "connected world."

Opportunities: More subjects will be enabled to access trials and will be empowered to participate more actively as data gatherers.

Threats: Neglecting to address regulatory issues early and failure of regulators to organize for the challenges will delay and possibly confound this opportunity. Covid-19 has demonstrated the cost of neglecting to embrace these ideas, first mooted in the First Edition of this book.

Take-home message

Empowering patients to collect their own trial data, more or less in the role of sub-investigators, offers an increasing opportunity to gather rich primary electronic source data directly, thereby displacing some of the traditional activities in trial sites for many trials. This is most obvious in chronic diseases. It should cause us to reevaluate what and how data are gathered in trials. We should start the management of change toward centralized monitoring of subjects, as well as of quality compliance. We can minimize the need for on-site monitoring and maximize business continuity during crises. We can greatly cut the cost of irrelevant activity and data that do not contribute to trial outcomes. We can start treating trial participants less as subjects and more as real stakeholders.

References

[1] O'Leary E, Seow H, Julian J, Levine M, Pond GR. Data collection in cancer clinical trials: too much of a good thing? Clin Trials 2013;10:624−32.

[2] U.S. Department of Health and Human Services Food and Drug Administration. Guidance for industry: electronic source data in clinical investigations. September 2013.

[3] European Medicines Agency. GCP Inspectors Working Group (GCP IWG). Reflection paper on expectations for electronic source data and data transcribed to electronic data collection tools in clinical trials. June 09, 2010. EMA/INS/GCP/454280/2010.

[4] European Medicines Agency. Notice to sponsors on validation and qualification of computerised systems used in clinical trials. April 07, 2020. EMA/INS/GCP/467532/201907.

Home nursing replacing site visits

Melissa Hawking, BS

Senior Manager, Marketing, Symphony Clinical Research, Vernon Hills, IL, United States

The need

Currently 86% of clinical trials fail [1]. Of the trials that fail, up to 85% are due to a lack of enrollment and the average patient dropout rate, which is 30% across all clinical trials [2]. So why is it so difficult to recruit patients to a clinical study and retain them? One reason is due to the time and travel burden placed on patients to participate. The 2017 CISCRP Perceptions and Insights study found that 60% of patients and caregivers listed the physical location of the research study center as very important in influencing their decision to participate in a trial [3]. The study further found that the location of the medical center was highly ranked as a least-liked aspect of patients' clinical study experience, second only to the possibility of receiving a placebo.

As a result, the need for innovation toward patient-centricity in clinical research has become increasingly apparent. These efforts have been supported by regulatory agencies, including the FDA. In January 2019, former FDA commissioner Scott Gottlieb expressed the FDA's support of decentralized clinical trials in an effort to make clinical research more "agile and efficient" [4]. Decentralized clinical trials are defined by the Clinical Trials Transformation Initiative (CTTI) as, "those executed through telemedicine, mobile, or local health care providers, using procedures that vary from the traditional clinical trial model" [5].

The solution

As drug companies and contract research organization's (CRO) move their trials toward a patient-centric approach, at-home trial visits have become more accepted, particularly during the COVID-19 pandemic. Taking some or all clinical trial visits out of a trial site and, instead, delivering them directly to patients in their homes can greatly benefit a clinical study overall.

- Trial continuity during times of crisis
  Like we saw during the COVID-19 pandemic, in times of crisis it may be very difficult for sites to continue seeing clinical trial subjects, or it may be more difficult for subjects to travel to sites for their visits. Having home care as part of

Innovation in Clinical Trial Methodologies. https://doi.org/10.1016/B978-0-12-824490-6.00019-0

your clinical trial from the beginning will make it much easier to continue trial visits at home without missing critical data points and without subjects missing drug doses that they may desperately need. We saw this in one case study example where a home care provider worked with a pharmaceutical company on a four-study program. The home care provider was already contracted for patient visits in year two of the study, but due to the COVID-19 pandemic, the sponsor requested that they move-up in-home services. This request was completed within 3−19 days for all four studies and no doses of study drug were missed.

- Decreased travel and time burden for patients
  Travel and time commitment can be huge barriers to a patient deciding to participate in a clinical trial, sometimes due to quality of life factors and sometimes due to a patient's inability to travel because of their health complications. In a 2016 survey, 54% of caregivers for patients with rare diseases reported that home-based research visits would increase their likelihood of research participation [6]. Offering some or all of a trial's visits in patients' homes can lead to higher patient satisfaction, accelerated recruitment, and increased retention.

- Reduced patient dropout
  As mentioned in the previous point, patients may be more motivated to stay in a study if their time and travel burden is reduced. However, home visits do not need to be constrained to the home. Home visits can also take place at a patient's workplace or school, and can even follow them on vacation, or if they need to travel for any reason. This service allows patients to remain in a study, who might have otherwise needed to dropout.

- Expanded geography
  By removing or decreasing the travel burden placed on patients, a single site can recruit patients from a larger geographic range, including those who may live out-of-state/province. This can lead to sponsors opening less sites for a trial, providing significant cost savings.

- Increased compliance
  When exploring virtual or decentralized trials, sometimes too much responsibility can fall to the patient to provide their own drug administration, collect their own lab samples, and record their observations. Using home clinicians in these cases can help the patient feel more connected to and supported by their clinical trial. Additionally, if compliance is an issue in a clinical trial, using home nurses to deliver or observe drug administration or to review patient diaries can be one way to increase compliance.

- Enhanced data collection
  For decentralized studies that are using monitoring devices, or relying on patient reported outcomes, adding in-home visits to the study can ensure that data is being captured regularly and accurately by a medical professional. In addition, some home nurses can use electronic data collection methods, which may be easier for the site to upload to their Trial Master File.

- Diversified intent-to-treat pool
  Home nurses can travel to patients' homes, wherever they are in the world. By not limiting recruitment to only patients that live near a study site, there's an opportunity to increase diversity in a clinical trial's patient pool.
- Eased site burden
  Running a clinical trial requires a great deal of effort on the part of a clinical trial site. By taking some visits to patients directly in their homes, the number of visits conducted by a trial site can be reduced, while potentially increasing their patient enrollment and satisfaction. Ultimately, the PI is still responsible for overseeing each patient's care so the site will retain all patient data.

Where does homecare fit in a clinical trial?

In-home trial visits can be used with virtually any therapeutic area or patient population. It is most beneficial in studies that are long in duration or have frequent visits to ease the travel burden. It is also a helpful addition to studies struggling with enrollment or retention, as a rescue strategy. Regarding patient population, home care is beneficial to immobile patients, whether the mobility challenges come from age (the very old or very young) or disease state. It's also beneficial in rare disease trials where patients are more likely to be geographically remote. Home nurses can help to bring the trial to the patient wherever they are around the globe, rather than bringing the patient to a distant trial site. Additionally, in-home visits are beneficial for a relatively healthy patient population that may be busy with lifestyle obligations, including work, family, or travel.

In-home services typically utilize registered nurses (RNs) to visit patients in their homes, but can also use phlebotomists and, on rare occasions, physicians, midwives, or other practitioners. In general, in-home nurses can conduct any activity, simple or complex, within their licensure and that can be made mobile. Some examples include the following:

- Patient education and training
- Blood draw and processing
- Biological sample collection
- Study drug administration—oral, topical, injection, IV infusion, nasal
- Drug observation
- Body system assessments
- Vital signs collection
- Review patient questionnaires or diaries
- Check for changes in health or medications; check for hospitalizations

Obstacles to implementing in-home clinical trial services

In-home nursing has been in the market as a clinical trial service for 15—20 years, but has only recently started gaining widespread acceptance. We've reviewed the benefits of utilizing in-home clinical trial services, but there are some challenges to implementing in-home services that need to be considered.

- Maintaining site relationships
  Some pharmaceutical companies worry that utilizing home care could negatively impact their relationships with study sites. This can be a concern in any decentralized clinical trial, as the goal is ultimately to bring the trial directly to the patient, which may reduce the number of sites needed for a study. It is important to note that expanding the geographic reach of trial sites via home visits can lead to better performance for the sites that are activated, allowing them to recruit and retain more patients. Additionally, home nursing not only reduces the patient burden, but the burden placed on site staff as well. It is often helpful to allow the principal investigator (PI) or site staff to meet the home care nurse assigned to their patients to ease any concerns about patient care. It is also vital to clearly define the roles of the PI, site support team, and the home care partner. Consistent communication and delegation of authority are important for a positive working relationship.
  In addition to a site's relationship with the sponsor or CRO, some worry that home visits may negatively affect a patient's relationship with their physician. In fact, offering home trial visits can decrease the frustration patients may feel with frequent or long site visits, therefore benefiting the patient/doctor relationship.
- Early inclusion in protocol and study design
  In-home visits need to follow many of the same regulatory restrictions as clinical trial sites. Therefore, home care needs to be addressed at the protocol level, requiring inclusion in the protocol and informed consent forms (ICF) and must be approved by any relevant institutional review boards (IRBs) or ethics committees. This requires some forethought on the part of the pharmaceutical company or CRO, or else, it will require protocol amendments to add home care down the road. Because of this, there can be a long lead or "start-up" time for introducing home care to a clinical study. In most cases, it may be best to include the option of home visits from the development of a clinical trial so that it will be easier to introduce later, if needed.
- Different regulations and customs globally
  When utilizing home visits, it is important to work with a partner who has an understanding of different home care regulations across the globe. For example, the nurse licensure will not allow the same services to be performed everywhere. In addition, they need to work within a country's culture around bringing clinicians into the home. Typically clinical trial home care partners work with local home nursing agencies that are well-versed in the local regulations and customs, and offer nurses who speak the local language.

- Cost

  Of course, adding home visits to a clinical study will correlate with an added cost. For pharmaceutical companies that have not used home care in the past, it can be difficult to allocate funds to innovative strategies that are unproven for them. However, the cost of home care can be offset by avoiding trial delays due to patient recruitment and retention. If the use of home care also allows for less sites to be opened, that can lead to cost savings as well.

- Reluctance to adopt new trial designs

  As with many decentralized trial tactics, there is some reluctance to move away from the traditional clinical trial model. However, as noted previously, the traditional model is expensive, time-consuming, and too often, trials fail due to a lack of patient enrollment. In-home trial visits can be a good place to start, because it is an easy tactic to test. For example, you can offer in-home visits only in certain countries, or only for specific visits within a trial. Or patients can be divided into groups—those offered home care and those offered traditional site visits. If a pharmaceutical company finds that offering in-home visits is beneficial to the trial, they can always expand the offering. Additionally, regulatory authorities, such as the FDA and EMA are accepting of in-home clinical trial services, so testing this decentralized service-model is relatively low-risk to the overall health of a clinical trial.

SWOT analysis

Strengths: Enhanced enrollment of a more naturalistic patient population with less dropouts. Provision of IP and collection of most safety assessments is also doable in crisis situations.

Weaknesses: Initial additional costs for further study setup requirements. Not all efficacy assessments are doable, but in conjunction with telemedicine that can get addressed in many instances.

Opportunities: Make most use of Patient Reported Outcomes at patient's home, thus providing more datapoints with overall less variability.

Threats: In a worst-case pandemic scenario, nurses may have trouble traveling to patients homes due to travel restrictions.

Take-home message

It is clear that the clinical research industry faces many hurdles with rising cost and time to develop a drug and the high number of trials that fail. We know that one of the main reasons for trial failure is a lack of patient enrollment, paired with patient dropout rates, and that this is often due to the heavy burden placed on patients. One way to lighten that burden is reduce the time and travel obligation for participating in a clinical trial, which can be achieved through in-home trial services. While we have explored the benefits and challenges to in-home clinical research, there are some recommendations for integrating in-home clinical trial services to an organization that's new to this service offering.

1. Include in-home services early in the trial design process, even if you are not sure you will use the service. This will allow for the option to more quickly implement the service if it is needed to bump patient enrollment, or for trial continuity purposes.

2. Keep open communication with your home care partner throughout the duration of their service. This will allow for any issues to be addressed and resolved quickly. Your partner's expertise can also be helpful throughout the trial, including the best way to include home care in the protocol and other documentation, or the best way to use home care alongside other decentralized trial tactics, such as remote monitoring, telemedicine, etc.

3. Start slow if you are having trouble with leadership buy-in. Home care can easily be structured into an "arm a/arm b" test and can be ramped up or pulled back easily depending on the results you are seeing in any particular trial.

References

[1] Hale C. New MIT study puts clinical research success rate at 14 percent. February 4, 2019. Retrieved from: https://www.centerwatch.com.

[2] Fassbender M. For clinical trials, 'convenience' services to be a standard offering in 2019. January 11, 2019. Retrieved from: https://www.outsourcing-pharma.com.

[3] CISCRP. 2017 perceptions & insights study. 2017. Retrieved from: http://www.ciscrp.org.

[4] Gottlieb S. Breaking down barriers between clinical trials and clinical care: incorporating real world evidence into regulatory decision making. January 28, 2019. Retrieved from: http://www.fda.gov.

[5] Clinical Trials Transformation Initiative. Decentralized clinical trials. Published. September 2018. https://www.ctti-clinicaltrials.org/projects/decentralized-clinical-trials.

[6] Amengual T, Adams H, Mink J, Augustine E. Rare disease clinical research: caregivers' perspectives on barriers and solutions for clinical research participation. February 8, 2016. Retrieved from: https://n.neurology.org.

Virtual visits: moving clinical trials visits from clinics to homes

John Reites
President, THREAD, Cary, NC, United States

Introduction

A Decentralized Clinical Trial (DCT) is defined as a clinical trial conducted using telemedicine, mobile, or local healthcare providers who are not limited in practice geographically, by the use of mobile technologies for increased frequency, or continuous data collection providing for a holistic picture of the patient rather than a single clinic visit snapshot in time [1].

Decentralized studies include technology and service approaches that enable remote data capture with patients, sites, home health, call centers, and other study stakeholders. Technology features of decentralized studies include the combination of the following:

- eConsent
- eCOA (ePRO, ClinRO, etc.)
- sensors (i.e., medical devices and consumer wearables)
- ediaries/surveys
- patient engagement solutions (i.e. content and reminders)
- patient authentication solutions
- Electronic Device Reported Outcomes (eDROs)
- eSource forms
- case report forms (CRFs)
- telehealth

Decentralized approaches enable global studies to move specific assessments and visits from completion in the clinic to the home. This hybrid decentralized approach shifts protocol schedule of events to appear as noted in Fig. 8.1 below.

Decentralized approaches enable global studies to move specific assessments and visits from completion in the clinic to the home. This industry-wide shift to decentralized studies is providing sponsors with the opportunity to enable their global clinical trials to be more;

- Mobile and flexible in their design
- Cost effective (depending on how decentralized a study is designed)
- Digital — capturing more data via eSource

**Capture data in-between visits and replace
specific in-clinic sessions with Virtual Visits**

FIGURE 8.1

Decentralized study schedule of events visualization.

- Convenient and supporting of participants lifestyles
- Responsive with instant options for telehealth sessions
- Expansive in data collection of new and continuous data in between visits

This shift has also been amplified in 2020 due to the COVID-19 pandemic. As a result, both active and upcoming clinical trials can utilize DCT approaches (as noted below) to minimize the risk of missing visits and assessments due to the pandemic's impact.

Active clinical trials

- Add eConsent (BYOD and/or provisioned device)
- Add on-demand telehealth virtual visit options
- Add flexible scheduling allowing visits to be conducted via telehealth, home health, and/or on-site
- Add source document collection and management to support data capture during virtual visits
- Move on-site assessments to technology-enabled activities

Upcoming clinical trials

- Virtual visits noted within protocols, schedule of events (SoE), and ICFs
- Modified physical exams and vital sensors included
- Allowing some visits to be only on-site/telehealth/home health and some visits with flexible option
- Mobile eConsent, eCOA/ePRO, surveys, reminders, sensors, eSource, telehealth, etc. becoming protocol standards
- Determining what assessments are unable to be completed remotely to influence SoE design

Given the complexity of clinical research and the desire to ensure research sites are engaged in the study conduct, implementing a fully decentralized study may not be an option for most studies. A hybrid decentralized study with a mix of remote data capture, in-clinic and virtual visits may fit the majority of specific study protocols best noted in Fig. 8.2.

Consider Your Study Design Goals	Hybrid Decentralized Study	Fully Decentralized Study
Increase operational efficiencies and participant convenience	X	X
Use in trial phases Ib – IV	X	X
Only utilize specific remote research solutions such as: eConsent, eCOA, ePRO, sensors, telehealth, surveys, etc.	X	
Develop a protocol that enables a fully decentralized study schedule of events		X
Recruit participants from a larger geographic area around sites	X	X
Requires no physical visits to the study clinic		X
Reduce overall budget by decreasing in-clinic visits, data reviews and source data verification	X	X
High clinician interaction requirements	X	
No hands-on clinician interaction requirements		X
Which remote research approach best aligns with your study goals?	A Hybrid decentralized study allows you to replace **some** in-clinic visits with Virtual Visits, collecting data during, in-between and in lieu of in-clinic visits. This approach minimizes the need for organizational change and enhances participant convenience.	A Fully decentralized study requires replacing **ALL** in-clinic visits with Virtual Visits. This approach requires significant changes to the protocol schedule of events and may require significant organizational changes. It may also provide the most participant convenience.

FIGURE 8.2

Differences between hybrid and fully decentralized study models.

What is a virtual visit?

A Virtual Visit is when a site team member(s) and study participant meet via a global regulatory compliant technology platform to conduct a telehealth video call and simultaneously complete study-required activities together during the session. Virtual visits allow the investigator to observe the participant, instruct them to complete assessments and record the site observations directly into the data capture system. Virtual visits can be conducted by study participants via scheduled sessions, on-demand sessions (i.e., unscheduled visits), and flexible sessions (i.e., when a participant is offered a choice of on-site or virtual visit for a specific study visit). Please see Fig. 8.3 for an example of virtual visit technology.

Site Team Experience
Video and audio call where a "visit" can
be conducted and data captured via
eConsent, eSource, eDC and/or
eCOA/ClinRO

Patient Experience
Video and audio call where a "visit"
can be conducted and data captured
via eConsent, activities, surveys,
sensors and/or eCOA/ePRO

FIGURE 8.3

Example of a virtual visit technology.

The key aspect of a virtual visit is that it enables activities to be conducted by the site and participant in real-time, with the validated approach during the active telehealth session. This process enables visit activities to be completed in the same manner (or as similar as possible) to how they would be performed in the clinic (Fig. 8.4).

How are virtual visits incorporated into protocols?

As noted in the Fig. 8.5 example, virtual visits support protocols to capture data via on-site and/or remote approaches. These approaches can be tailored to each protocol based on a number of factors including, but not limited to, the complexity of the protocol-required assessment, validation requirements of a specific assessment, patient population, if home health or call centers are included, etc.

Step #1
Site Team logs into a portal on
any device with internet browser
at scheduled time and clicks to
start a telehealth Virtual Visit

Step #2
Patient receives notification on
BYOD or provisioned
phone/tablet, accepts call with one
click and starts the Virtual Visit

Step #3
Site Team completes forms and
asks patient to complete
activities on app during the
Virtual Visit

FIGURE 8.4

High-level virtual visit process.

Assessment	Data Capture Technology	Who Contributes the Data		Where is the Data Captured	
		LITE Decentralized	EXPANDED Decentralized	LITE Decentralized	EXPANDED Decentralized
Informed Consent	Platform	Site and Patient on Paper	Site and Patient	On-site	On-site
Demographics and Med History	EDC	Site	Site	On-site	On-site
eCOA / ClinRO	EDC + ClinRO	Site	Site / Rater	On-site	Remote
Prior Therapy	EDC and/or eSource	Site	Home Health	On-site	Remote
Physical Exam	EDC and/or eSource	Site	Home Health	On-site	Remote
Vital Signs	EDC and/or eSource	Site	Home Health	On-site	Remote
Labs	EDC and/or eSource	Site	Home Health	On-site	Remote
ECG	EDC and/or eSource	Site	Home Health	On-site	Remote
Patient Reported Outcomes	Patient app with ePRO included	Patient self-reported	Patient self-reported	Remote	Remote
Administration of Study Drug	EDC and/or eSource	On-site w/ additional supplies provided	Shipment via Home Health/DtP Shipping	On-site w/ supplies provided	Shipment via Home Health/DtP Shipping
Sensor (Activity Tracking)	Patient App with medical device / sensor	Patient self-reported	Patient self-reported	Patient self-reported	Patient self-reported
eCOA / ePRO	Patient App	Patient self-reported	Patient self-reported	Patient self-reported	Patient self-reported
eDiary	Patient App with eDiary	Patient self-reported	Patient self-reported	Patient self-reported	Patient self-reported

FIGURE 8.5

Example schedule of events map.

Depending on the type of assessment and/or its validation requirements, specific activities can be conducted during virtual visits with telehealth and/or home health to ensure quality of the data captured.

Virtual visit benefits

There are a number of benefits to virtual visits beyond being able to conduct sessions remotely outside of the clinic. A few of the key advantages are noted below:

- Connecting with participants is important and key to engagement, data collection and retention. Conducting these sessions in person with a video connection allows for a more personal connection. People want to interact, especially with healthcare providers. Seeing them live supports their confidence and comfortability to continue to participate in the study, lending a deeper engagement with the participant
- Ability to confirm the patient is who they say they are via ID verification over video
- Ability to consent or reconsent via the platform during a virtual visit
- Ability to view data and show reports via camera and/or screen sharing
- Ability to include multiple site team members from remote locations into a single session
- Broadcast study updates directly to participant through the app
- Site staff can access using most any web browser. This is important as many site staff are restricted from office access at this time, thus the staff can communicate directly with the participant from their home as opposed to having to rely on a tablet kept at the site or phone numbers stored on office equipment.

Additional benefits by study stakeholder may include the following:

Participants
- Reduced travel, time on-site, and administrative time
- Ability to participate in study from rural area with mixed support
- Telehealth option for unscheduled/on-demand sessions
- Technology available to engage data, contribute data, and conduct visits
- Value and support specific to the patient population
- Reminders/notifications to keep to the study schedule
- One-click support to call or email helpdesk

Sites
- Continuous view of patient data and similar process to on-site visit
- Decrease on-site visits and increase interaction
- Identify "at risk" patients and/or safety alerts in between on-site visits
- Reduce administrative burden when technology is alleviating, not adding to workload
- Provides an on-demand unscheduled visit capability to decrease travel/administration
- Can work directly with home health services

Sponsors/Clinical research organizations (CROs)
- Modern research experience for patients and sites
- Reduced research costs
- Potential recruitment geography expansion based on how hybrid the virtual design is
- Continuous data collection
- Potentially more frequent safety review
- New endpoints, measures, and data context
- Differentiated support tools for the study patients and sites

Virtual visit challenges
Virtual visit approaches also have challenges that must be addressed for each study and may include the following by study stakeholder:

Participants
- Patient safety management process can be more complex
- Continuous data review required

- Less face-to-face interaction—mixed approach may not be best for everyone
- Requires technology backup plans
- If options not flexible, may increase burden

Sites

- Technology learning curve
- New training required for site and home health team members
- More data being collected
- New and/or revised SOPs may be required
- New solutions for study teams to learn
- Shift in research business model

Sponsors/CROs

- RA/Agency discussions require context and a clarity on the decentralized approach planned
- Design based on the validated assessments
- Requires a backup plan for technology
- New and/or update SOPs may be required
- New solutions for study teams to learn

Conclusion

Virtual visit approaches enable global studies to move specific assessments and visits from completion in the clinic to the home. When conducted in hybrid decentralized study approaches, virtual visits have the ability to provide a more modern, flexible study experience for all research stakeholders.

Although this is not a new concept, this global industry shift to virtual visits has seen acceleration in adoption from the impact of the COVID-19 pandemic. As we look forward to a postpandemic world, our societies recently formed virtual habits coupled with more flexible, technology-enabled approaches will become a new standard in clinical trials.

Reference

[1] Apostolaros M, Babaian D, Corneli A, et al. Legal, regulatory, and practical issues to consider when adopting decentralized clinical trials: recommendations from the clinical trials transformation initiative. Ther Innov Regul Sci 2019. https://doi.org/10.1007/s43441-019-00006-4.

Digitalized planning - digitalized studies

Data mining for better protocols

9

Fareed Mehlem

Medidata, a Dassault Systemes company, New York, NY, United States

The need

Good protocol design is critical to the success of a clinical program. It must balance needs across scientific, commercial and regulatory outcomes, and it must be feasible to execute from an operational perspective. Furthermore, the COVID-19 pandemic has heightened the need for thoughtful design that minimizes patient burden while maintaining safety and data quality.

The impact of protocol design can be massive. Over 70% of trials fail to meet their target timelines[1], often driven by overly restrictive or complex protocols. An even more direct and damaging outcome of poor trial design is the need for protocol amendments. According to a 2016 study by the Tufts Center for the Study of Drug Development, 57% of protocols have at least one significant amendment, resulting in delays of 3+ months, and estimates put the median direct cost of a phase III amendment at $535,000 not including the indirect cost of lost time to market[2].

And the challenges involved in protocol design are growing:

- Products are becoming more specialized and complex, with growth in combination regimens, gene therapies, and other new modalities.
- Trials are targeting more specific patient populations with the growth of rare disease and the move toward precision medicine.
- Regulators are accepting innovation in trials including adaptive trials, virtualization of trials, and inclusion of Real-World Evidence (RWE) or Synthetic Control Arm (SCA) data.
- During the pandemic limiting the number of patient-site interactions became a new "must" and the use of virtual/remote technologies grew.

---

[1] Accelerating study start-up: the key to avoiding trial delays: ACRP.
[2] Getz K.A., Stergiopoulos S., Short M., Surgeon L., Krauss R., Pretorius S., Desmond J., Dunn D. 2016. The impact of protocol amendments on clinical trial performance and cost. Therapeutic Innovation and Regulatory Science. 50:436—441.

Innovation in Clinical Trial Methodologies. https://doi.org/10.1016/B978-0-12-824490-6.00021-9

In the face of these challenges, we have seen protocols become more complex over the last 10 years, including a 70% increase in total number of procedures per protocol and a 25% increase in planned visits[3] This translates to significant additional burden for sites and for patients, ultimately becoming unmanageable in the pandemic.

At the same time, pipelines are increasingly crowded especially in TAs such as oncology, immunology, and rare disease, driving intense competition for patients and sites. In a hyper-competitive trial landscape, this increase in protocol complexity can be a driving factor in enrollment and overall success. Increasingly, pharma companies are looking to reduce site burden and patient burden as a lever to drive competitive advantage in the search for patients.

For one, the fundamental way that study teams operate must change. Historically, the protocol has been written from the clinical point of view and then passed to operations teams downstream to implement. While most sponsors have mechanisms for operational input into the design, this often does not happen until later in the process, limiting the degrees of freedom in optimization. Part of the value of a data driven approach to design is that it provides a more objective view of tradeoffs between objectives (both core and exploratory) and operational impact. These tradeoffs discussions should move upstream early in protocol design process to drive true optimization and balance.

Additionally, there is opportunity to greatly increase the quality and amount of protocol data available for analysis by using structured tools to develop protocols in the first place. Today, the process of extracting data from protocols is time consuming and labor intensive. The majority of this data is unstructured, locked in formats such as PDFs, Word docs, and paper. There has been some success with the application of Natural Language Processing (NLP) techniques to automate the extraction of portions of protocols; however, protocols are not written in a standard way or with standard nomenclature across or even within sponsors. This limits the ability of NLP to be deployed to fully automate the extraction of data from protocols. Instead, full extraction requires teams of clinical experts to manually review and log data. Because of this, only a small portion of the data available in the world's protocols is currently available for analysis.

The solution

Data-driven protocol design can help mitigate these challenges by improving both scientific and operational design. Real-world and trial data can be used to significantly improve the scientific design of trials ensuring that the study is targeting a viable pool of patients and is appropriately powered. Historic protocol data can support better operational protocol design, driving alignment between procedures and objectives while minimizing patient burden.

---

[3] Getz and Campo, "New benchmarks characterizing growth in protocol design complexity", Therapeutic Innovation and Regulatory Science, June 2017.

Improving scientific design

Several questions must be addressed when designing trials:

- **Endpoints:** What endpoint(s) should be used and how should it be measured?
- **Study arms and comparators:** How many study arms should be included and what comparators should be used (placebo vs. active control)?
- **Dosing:** What are the level, route, and frequency of dosing?
- **Inclusion/exclusion criteria:** What patients are to be included and excluded from the trial? What are the disease parameters, demographics, medical history, comorbidities, and concomitant medications allowed and excluded?
- **Power:** How many subjects are required in the trial to properly power the study?

A number of sources of information are traditionally used to make these decisions including review of literature, internal experience and program knowledge, and expert input from investigators and key opinion leaders. These sources can be powerful but are limited by their inherent bias—they can only report on what they know.

Real-world data and clinical trial data can be used to supplement these sources and provide a quantitative analysis of patient profiles and past trial behavior. Real-world data is especially useful in defining patient populations to understand the impact of inclusion/exclusion criteria on available patients. Real-world data can also be used to support endpoint definition, dosing, and power by examining class effects versus standard of care.

However, for these analyses, clinical data provides several advantages over real-world data:

- Endpoints and covariates traditionally used in clinical trials are present and measured in the usual way in these data. This is in contrast to real-world data where these variables need to be created from related concepts.
- Clinical-trial data are high quality data generally with careful and structured data collection that is monitored and locked.
- Clinical-trial participants may differ from the larger disease population by socioeconomical or other factors and so these data are the most relevant data for understanding expected outcomes in clinical trials.
- Clinical-trial data tend to have a broader geographic distribution that commonly used real-world datasets, which often cover a single country or region.

In practice, it is useful to bring both real-world data and historic clinical-trial data together to better profile patient populations and disease characteristics. By looking at these data side-by-side, it is possible to understand the differences in real-world versus trial populations (see Fig. 9.1).

These data can help drive better decisions around trial design by allowing a more precise profile of patient characteristics.

Reducing operational complexity—site and patient burden

When study teams are designing protocols, they must consider the feasibility of the protocol in terms of operational complexity of the design, measured as site burden

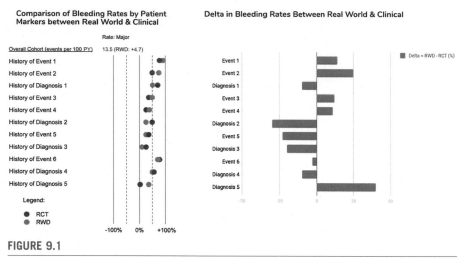

FIGURE 9.1

Comparison of real-world data versus clinical-trial data.

and patient burden. Overly burdensome protocols can lead sites to direct patients to more favorable studies, and patient burden can also have downstream effects—dropout rates in trials can be substantial, often 30% or higher.

Historically, study teams have used primarily qualitative measures to assess protocol burden, relying on conversations with investigators and site staff, spot checks versus past protocols, and personal experience. However, more objective approaches to assessing site and patient burden are increasingly being deployed. By deconstructing the schedule of events in a protocol, and benchmarking against a database of industry standards, a study team can objectively look at a protocol, understand how procedures align to objectives, how much they cost, and how consistent or unique the procedures are compared to other protocols in the indication (see Fig. 9.2).

In addition, it is possible to objectively measure how burdensome a protocol is for sites and patients. This is done by scoring every procedure for burden, and using that to build up a view across a trial. For example, Medidata has developed the Patient Burden Index (PBI), which translates detailed procedure and activity data into quantifiable, objective measures of patient burden. These measures were built and validated in partnership with the Society of Clinical Research Sites (SCRS) through a survey of investigator site personnel and patients going through clinical trials. The PBI includes quantitative measures across eight components to assess the toll a trial takes on a patient physically, mentally, and in terms of time. The components take into account the burden of procedures, including pain, invasiveness, harmful exposure, hospitalization, and anxiety, as well as the burden of

Activity Benchmarks

Objective Type	Endpoint Type	Code	Activities*	Study Activity Quantity			Benchmark						Patient Burden Index
				Min	Expected	Max	Activity Average Quantity	Activity Quantity Range	Visit Average Quantity	Protocol Usage (%)	Median US Cost	Protocol Complexity	
Sc, Tr		99211	Brief Visit w/ Vitals	2.00	2.00	2.00	7	1-11	23	67%	$20	0.18	1.50
St, E	KS-St, E-Q	82575	Creatinine, Clearance (GFR)	8.00	8.00	8.00	1	1-1	13	11%	$28	0.15	1.50
E	E-Q	81002	Dip Stick UA w/o Microscopy	4.00	4.00	4.00	15	6-23	20	89%	$25	0.04	1.50
S	S-Ef	85025	Hemogram (CBC) w/ Plate & Auto Diff	1.00	1.00	1.00	18	9-37	19	100%	$35	0.11	1.50
P, S, Sc	KS-St, S-Ef	87340	Hepatitis B Surface Antigen (HBsAg)	5.00	5.00	5.00	1	1-1	19	56%	$110	0.11	1.50
U		*INEX	Inclusion/Exclusion Criteria	2.00	2.00	2.00	1	1-2	19	100%	$100	0.20	1.50
P, Or, O	P-Ef	99205	Initial Visit w Hist, Phys & Vitals	4.00	4.50	5.00	1	1-1	19	100%	$200	3.17	1.50
O, O	O-PE	85730	Partial Thromboplastin Time (PTT)	5.00	5.00	5.00	9	5-13	17	67%	$30	0.06	1.50
E, Sc, Tr, O	E-Q	83520	Quantitative Immunoassay	8.00	8.00	8.00	8	8-8	24	22%	$15	0.11	1.50
P	KS-St	84702	Serum Pregnancy Test, Quantitative	1.00	1.50	2.00	1	1-1	25	22%	$53	0.22	1.50
S, O	T-St, O-PE	80299	Single Drug Level/PK; Any Source [2]	9.00	9.00	9.00	19	7-57	19	100%	$50	0.18	1.50
E, Sc	E-Q	NC125	SMAC 19: 13+ Chemistries [2]	10.00	10.50	11.00	19	9-36	19	100%	$25	0.33	1.50
E	E-Q	N84156	Total Urine Protein [2]	6.00	6.00	6.00	N/A	N/A	N/A	0%	$18	0.06	1.50
Tr		84520	Urea Nitrogen [2]	1.00	1.00	1.00	N/A	N/A	N/A	0%	$25	0.06	1.50

Study metrics include activities and visits that are marked optional/conditional.

2 Collection cost only. Central lab cost not included.

Study Activity Quantity Min out of Benchmark Activity Quantity Range   OR   Protocol Usage ≤ 10%

Study Activity Quantity Min above Benchmark Activity Average Quantity   OR   Protocol Usage ≤ 33%

N/A   Data not available for current specificity

FIGURE 9.2

Study activity burden scoring and benchmarking[4]

questionnaires and assessments in terms of time, items, and type[5] These data can be used to support patient-centric protocol design, improving patient experience and reducing the risk of poor subject recruitment and patient dropout.

Site and patient burden should be optimized not only for the overall protocol, but across the trial journey, day by day and visit by visit. By mapping burden to the visit schedule, a sponsor or Contract Research Organization (CRO) can better balance scientific need and burden, and better manage expectations of sites and patients. In the example below (Fig. 9.3), you can look across a trial and see spikes and troughs. Here we see a spike in terms of patient intensity at the Day 120 visit. This is a real example—in this case the sponsor looked at this visit and realized that it included several high-burden procedures that supported exploratory objectives. To mitigate the risk of patient dropout at Day 120, the sponsor made a strategic decision to remove some of these procedures from the visit, reducing the burden without sacrificing core objectives.

---

[4] Screenshot of Medidata's Study Design Optimizer solution.
[5] https://www.medidata.com/wp-content/uploads/2018/10/Using-Patient-Burden_White-Paper_201805_Medidata.pdf.

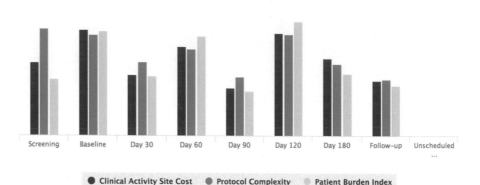

FIGURE 9.3.

Example of cost and burden by visit.

Take-home message

Protocols are becoming more complex and driving significant negative impact on clinical programs. Over 70% of trials fail to meet their target timelines and 57% of protocols have at least one significant amendment. In the pandemic situation, complex protocols with frequent site-patient interactions became undoable. Data-driven protocol design can help mitigate these challenges by improving both scientific and operational design.

- Leverage real-world and clinical-trial data to define endpoints, dosing, inclusion/exclusion criteria, and power.
- Leverage detailed historic protocol data to optimize operational design and minimize site and patient burden.
- Integrate both of these approaches early in the protocol design process to shrink timelines and explicitly discuss tradeoffs.
- The industry should move toward structured protocol design, using authoring tools to construct protocols in a standard way and automating the collection of data from these structured protocols. This will help solve the downstream data availability problem, and also create a more efficient process for the design of protocols and the flow of that information into downstream processes (e.g., eCRFs).

Patient-powered registries for population enrichment

10

Peter L. Kolominsky-Rabas, MD, PhD, MBA

Head, Interdisciplinary Centre for Health Technology Assessment (HTA) and Public Health,
Friedrich-Alexander-University, Erlangen, Germany

The need

Even though novel concepts may limit the need for the number of patients enrolled in a given clinical trial (see Part 2: Alternative Study Concepts Requiring Less patients), patient enrollment is still and will remain a key rate-limiting and cost-driving step in clinical research.

That is why the idea came up to create so-called "trial-ready" cohorts through upfront registries. This idea had major impact on clinical research in Alzheimer to simplify the access to patients in early disease stages. As example, see Ref. [1].

In Alzheimer disease (AD), a recent survey of Karolina Krysinska et al. [2] revealed 31 ongoing AD registries across the globe. More than half of the registries aimed to conduct or facilitate research, including preclinical research registries and registries recruiting research volunteers. In the USA, five registries were identified which were exclusively built to facilitate the enrollment of prodromal or early stages of the disease [3].

Not only studies in Alzheimer would benefit from patient populations, which are already well defined and in parallel expressed their interest to participate in a trial.

However, in other diseases, registries are not yet as common as in AD. In 2015, the EU Joint Programme Neurodegenerative Disease Research (JPND) reviewed the registries in AD, frontotemporal dementia, Huntington disease, Parkinson disease, prion disease, motor neuron disease, and the spinocerebellar ataxias. In all entities only local registries were in place, except in Parkinson. This creates the questions whether and under what conditions it would be worth an investment in patient registries, also in other disease entities. This is an even more relevant question, since we better and better understand the factors which describe the response to treatment (see Fig. 10.1). Knowing such factors before a patient gets invited for a screening visit would make the screening process much easier and cheaper.

A second question is related: How to motivate individuals to register and (what is even more challenging) to regularly update their registry-information while they are not yet enrolled in a study? This question is critical, since we know that only about 6% of patients with a severe disease participate in clinical trials [4].

Innovation in Clinical Trial Methodologies. https://doi.org/10.1016/B978-0-12-824490-6.00010-4

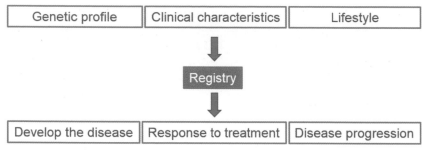

FIGURE 10.1

Nowadays registries can capture multiple modalities: genetic profiles, a variety of biomarkers in combination with clinical data and insights in the individual's lifestyle, provided through social media and wearable data. This allows a more precise prediction of the individual's prognosis, beginning with the risk of suffering the disease in near future to the responsiveness to treatments and future natural course of the disease.

The solution

Registries are defined as "an organized system that uses observational study methods to collect uniform data (clinical and other) to evaluate specified outcomes for a population defined by a particular disease, condition, or exposure, and that serves a predetermined scientific, clinical, or policy purpose(s)". [5,7].

Various guidance exists for the proper set-up of a registry [5,6,8,9].

According to Workman TA [10], one should mainly differentiate between the two types of research-centric databases:

- Population-focused/-based registries (enrolling a more representative national or regional sample of the population, also often named a "cohort"). These are often closely linked to the national public healthcare system and re-imbursement, such as in the Swedish SveDem which reached a national coverage rate of enrolling 30% of newly diagnosed patients with AD in 2012. Another example where the reporting of patients in a registry is a mandatory requirement is the California Parkinson's Disease Registry [www.capdregistry.org]. Also patient-empowered registries are most typically population-focused, such as the Michael J Fox trial-finder registry, also in Parkinson disease.
- Hospital-based databases. These are often more selective, thus potentially having to some degree a selection bias.

However, when managed by researchers, the registry may provide little or no opportunity for involvement or control by patient or family members or patient support and advocacy organizations. As a result, the registries may not meet the needs of patients, family members, and informal caregivers, as well as advocacy groups [10,11].

Logically, that would limit the ability of these registries to be of interest for the targeted population. The alternative is:

1. Patient-powered registries [6]. These are in many ways similar to researcher-generated patient registries, with one exception: patients and family

members, not researchers, "power" the registry by managing or controlling the collection of the data, the research agenda or the data, and/or the translation and dissemination of the research from the data. An effort to document such patient-generated registries is being undertaken by the American Association for the Advancement of Science through funding by the Agency for Healthcare Research and Quality [12]. A study of 201 disease advocacy organizations found that 45 % had supported a research registry or a biobank [13].

This differentiation is relevant, since it is not only determining what group initiates the registry (academia vs. patient groups), but also to what objectives are fulfilled.

Academic, research-centric registries are mainly interested in:

A. Facilitating the recruitment of patients for clinical trials. K Johnson et al. [14] provide a good overview of what methods of enrollment in a disease registry work best. Direct mailing turned out to be most expensive but of limited effect, while paid internet advertisements (Facebook and Google) were most effective and yielded over 65% of all enrolled individuals. Indirect methods (word of mouth, Twitter, advocacy websites) were contributing only about 15%. The most promising method however depends on the age group, with those above 65 years still preferring classic direct mailing [15].
B. Epidemiology: Learn about population behavior and their association with disease development.

The patients are more interested in aspects such as:

A. Current information about their disease.
B. Support the development of new therapies.
C. Empowerment to better manage their chronic disease.

There is of course overlap of objectives between both groups. The key question: Which of these objectives are crucial to be addressed for the success of registries which shall facilitate enrollment in early disease treatment studies?

Individuals entering such registries do that in general if three conditions are fulfilled [15].

1. For altruistic reasons; patients expect their contribution may help others. Regular (e.g., bi-annually) status reports shared with all stakeholders (including all registered participants) are a good tool to ensure this aspect is fulfilled. An example for such a status update report is provided by the Cystic Fibrosis Foundation [16]. Another option to address that need is the ability of members to not only participate in top-down shared information, but also to communicate bottom-up or even horizontally with the peer group. The patient-powered registry "patientslikeme" fully implemented that requirement (see also Fig. 10.2).
2. Insights and access to new therapies; they expect their own care may benefit. This aspect is detailed by Ref. [17]. The result could be a win-win for both key stakeholders (healthcare providers and patients) by mutually providing status updates as outlined in Fig. 10.2.
3. Convenience; enrolling in a registry must be easy. For that purpose, a number of technical solutions are available, mainly to register online. This however

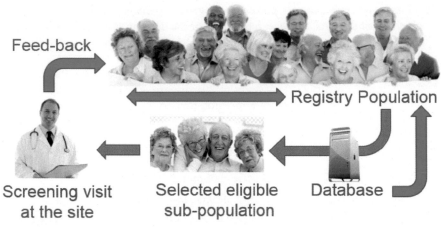

Feed-back

Registry Population

Screening visit at the site

Selected eligible sub-population

Database

FIGURE 10.2

Registries with a focus on being attractive for registrants should not only collect information from the individual into the database in a "one way street" approach (*gray arrows*). Patient-centric registries allow the retrieval of information from the data-repository plus the exchange of information among the peer-group of patients such as informal caregivers, family members, advocacy groups. In other words, the registry also fulfills the function of a disease-specific social medium. This function is enhanced by additional feedback from the academia, e.g., the participating study centers (*green arrows*).

creates the risk of duplicate or even falsified entries, what led to the concern that "only a small minority of patients with sufficient education and ability are able to participate, and that data may be biased" [18]. However, for registries exclusively focusing on the pre-identification of individuals at risk, this is much less an issue than for registries which shall generate epidemiologic data or even support a drug registration. Also, limiting competition and reducing the fracturing of efforts to collect data, raise funds, or advance knowledge is a relevant aspect of convenience. However, until such "best practices" registries are not yet fully in place, competition for best solutions should not be restricted.

In such shared models, a fourth aspect is of relevance, more in Europe than in United States: data privacy—makes sure any data provided are only used for the detailed purpose and otherwise safe.

Another item that should be kept in mind when setting up a registry is the appropriate data structure. Such a default data structure is detailed in Ref. [19]. For a patient-powered registry we propose the following most lean structure:

Basic data

- The consent for the registry and data use (by patient and legal guardian; of growing relevance is the upfront provided consent while the patient is still capable of consenting for activities, e.g., participation in trial or in any bio-sampling, what may only happen after disease progression)

- Patient diagnosis
- Patient personal information (name, gender, date, and place of birth)
- Disease history (age at onset and at diagnosis)

  Epidemiology

- Familial or sporadic
- Genes identified (if any)

  Clinical status

- Treatment
- Symptoms

  Research status

- Agreement to be contacted for a trial
- Agreement for specific procedures (CSF sampling, biosampling, MRI, PET)
- Having already given a biologic sample

Since not all registries follow a consistent format, it leaves the weakness that an aligned approach which would use all existing registries in a similar fashion is not possible. However, in large clinical trials in AD, targeting the enrollment of some-times over 1000 participants in phase 3 studies [20], a harmonized approach would make contacts more efficient.

A fully compatible data structure may however become less and less relevant due to the improving abilities to perform data-mining in various databases [21].

With IT technology progressing, not only various registries may get linked, but also registries and biobanks and/or Electronic Health Records (EHR). EHRs may thus supplement the current role of registries, even though they usually do not allow direct contact with an individual patient. Such links into other database formats would also open doors to the use of Patient Reported Outcome Data in registries. More and more patients use wearable technology through smartphones. These may provide very conveniently captured information about disease onset and pro-gression, as already done in Parkinson [22] or in Alzheimer [23].

The implementation and maintenance of a registry which fulfills all objectives may support enrollment for a planned large clinical development program, but it re-quires early upfront planning and is a major investment. For instance, in Australia, the cost of establishing a major national registry (50,000 cases reported annually), including the cost of the IT systems, has been estimated in 2013 at approximately US$ 0.5—US$0.75 million and the annual cost of maintaining such a registry at approximately US$0.75—US$1 million [24]. Such an investment to enhance enroll-ment in a clinical development program should provide an appropriate return.

Patientslikeme.com managed to become a profitable registry with over 600,000 members in 2800 different conditions. That may be due to the fact this registry is built bottom-up, i.e., a patient-powered registry which applies best practices to become an attractive and user-friendly tool that is not only collecting information but also offering ways to individually retrieve information from the system. It may thus be an alternative to building your own registry to check whether yet existing registries may also contribute to a planned program.

The concern around such "bottom-up" built registries is that only a minority of patients with sufficient education are able to participate, and that data may thus be biased.

Furthermore, when collecting data from patients online, there is the distinct possibility that users are not who they appear to be. In registries which have the objectives B (Epidemiology), C (Learn about the disease), or D (Support the development of new therapies), it is crucial to have access to reliable data. If the only objective is A: Enhance enrollment, this restriction is less an issue, since fake entries would not respond to any request for participation, while anyway, only a small portion of the registered individuals qualify for a given study.

Such fake entries nonetheless may increase administrative efforts for the registry and also create unnecessary high failures during initial screening for a new study.

As a solution, many websites ask users to enter minimal information about themselves, thus lowering the barrier for misrepresentation [25].

Combining a registry with a biobank would nearly eliminate that risk, since every registered person would have to provide a biosample—such as in the for-profit registry 23andme.com.

Last but not least, smartphone-based technology would allow remote identification through scanned and securely transmitted identity cards, a technology already used by banks and for airline check-in.

Thus, from the characteristics for the quality of a registry [9] only the below two items apply to the herein described Type A registries.

- A governance layer (e.g., it should have a charter and defined oversight roles, and a business plan, all being publicly available) is required.
- Data security for all registries capturing any type of privacy information.

  Data quality and the information quality are less relevant (as detailed above).

SWOT analysis

Strength: Registries help to pre-identify "ready to enroll" cohorts and can thus limit the costs for screening and enrollment

Weakness: Not all indications are best suited for this approach. Rare diseases and those where populations "at risk" are targeted benefit most.

Opportunity: Registries should be built "bottom-up," primarily addressing patient needs, as otherwise patient acceptance, data quality, and return on investment will be low. In some situations, when combined with naturalistically collected Patient Reported Outcomes (also see sections: "The use of new digital endpoints" and "The patient as PI"), these simple large registry-based studies may replace more complex classic randomized trials.

Threat: In future, more easy-to-access and more data-rich electronic health records may replace registries.

Take-home message

Registries which primarily support the enrollment in trials can be of a quite simple design, which should allow a rapid and large-scale set-up with limited initial investment. High compliance and retention can get achieved with a "patient-powered" design that also addresses the need for feedback and interaction to/from and in-between the registered individuals. Modern IT technology should be applied to allow links into other similar registries and the integration of further data-sources, such as biobanks, patient-reported outcomes and data from patient's smartphone.

References

[1] ClinicalTrials.gov identifier: NCT04004767; also see; https://www.aptwebstudy.org/welcome.

[2] Krysinska K, et al. Dementia registries around the globe and their applications: a systematic review. Alzheimer Dement 2017;13:1031—47.

[3] Aisen P, et al. Registries and cohorts to accelerate early phase Alzheimer's trials. J Prev Alzheimers Dis 2016;3:68—74.

[4] Getz K. The gift of participation: a guide to making informed decisions about volunteering for a clinical trial. Bar Harbor, ME: Jerian Publishing; 2007.

[5] Gliklich RE, Dreyer NA. Registries for evaluating patient outcomes: a user's guide. AHRQ Publication No.10-EHC049. Rockville, MD: Agency for Healthcare Research and Quality; 2010.

[6] 21st century patient registries. In: Gliklich RE, Dreyer NA, Leavy MB, Christian JB, editors. EBook addendum to registries for evaluating patient outcomes: a user's guide. 3rd ed. Rockville, MD: Agency for Healthcare Research and Quality; February 2018. AHRQ Publication No. 17(18)-EHC013-EF.

[7] Zhang S, Gaiser S, Kolominsky-Rabas PL, National Leading-Edge Cluster Medical Technologies "Medical Valley EMN". Cardiac implant registries 2006—2016: a systematic review and summary of global experiences. BMJ Open 2018;12 (IF 2.73).

[8] Niederländer CS, Kriza C, Kolominsky-Rabas PL. Quality criteria for medical device registries: best practice approaches for improving patient safety — a systematic review of international experiences. Expert Rev Med Devic 2017;14(1):49—64 (IF 2.08).

[9] Zaletel M, et al. Methodological guidelines and recommendations for efficient and rational governance of patient registries. PARENT; 2015.

[10] Workman TA. Engaging patients in information sharing and data collection: the role of patient-powered registries and research networks. AHRQ Community Forum White Paper. AHRQ Publication No. 13-EHC124-EF. Rockville, MD: Agency for Healthcare Research and Quality; September 2013.

[11] Ulbrecht G, Gräßel E, Nickel F, Kolominsky-Rabas PL. Low-threshold consulting services for dementia. Nervenarzt 2018;10 (IF 0.7).

[12] Social networking and online health communities: identifying and describing patient-generated registries. 2012.

[13] Landy D, Brinich M, Colten M, et al. How disease advocacy organizations participate in clinical research: a survey of genetic organizations. Genet Med 2012;14(2):223—8.

[14] Johnson K, et al. Evaluation of participant recruitment methods to a rare disease online registry. AJ Med Gen 2014;164A:1686—94.

[15] Solomon DH. Determining the best methods for using patient registry data in clinical research. Patient-Centered Outcomes Research Institute; June 2018.

[16] Cystic Fibrosis Foundation. Annual data report technical summary. 2018. Bethesda.

[17] Nelson EC, et al. Patient focused registries can improve health care. BMJ 2016;354: i3319. https://doi.org/10.1136/bmj.i3319.

[18] Frydman G. Patient-driven research: rich opportunities and real risks. J Particip Med 2009;1(1):e12.

[19] EUCERD Joint Action. Minimum dataset for rare disease registries. 2015.

[20] Luo J, et al. Minimizing the sample sizes of clinical trials on preclinical and early symptomatic stage of Alzheimer's disease. J Prev Alzheimers Dis 2018;5(2):110—9.

[21] Sernadela P, et al. Linked registries: connecting rare diseases patient registries through a semantic web layer. BioMed Res Intl 2017. Article ID 8327980.

[22] Trister AD, Dorsey ER, Friend SH. Smartphones as new tools in the management and understanding of Parkinson's disease. npj Parkinson's Dis 2016;2:16006. https://doi.org/10.1038/npjparkd.2016.6.

[23] Mc Carthy M, Muehlhausen W, Schüler P. The case for using actigraphy generated sleep and activity endpoints. J Prev Alzheimers Dis 2016;3(3):173—6.

[24] Australian commission on safety and quality in Health Care. Framework for Australian quality registries. Sidney: ACSQHC; 2014.

[25] Frost J, Okun S, Vaughan T, et al. Patient reported outcomes as a source of evidence in off-label prescribing: analysis of data from PatientsLikeMe. J Med Internet Res 2011; 13(1):e6.

Further reading

[1] Schüler P, et al. Presentation at 10th CTAD conference. 2017. Boston.

How to make a protocol patient-centric?

11

Peter Schüler, MD

Senior Vice President, Drug Development Services Neurosciences, ICON, Langen, Germany

The need

Nowadays Clinical Trial Protocols are written in a fashion that was introduced with the first version of GCP—and didn't improve since then.

They need to fulfill a variety of objectives—and only some of these are incorporated in a typical protocol. Often still missing are the following objectives:

1. Identify potential study risks (scientific and operational) and proactively mitigate these through improved study design.
2. Give reasons for any requested activity or restriction, so site staff not only have to "follow orders" but also understand the rationale.
3. Limit the study complexity as much as possible by reducing the number of secondary endpoints, subject selection criteria, assessments, and visits.
4. Involve potential trial subjects in the design, so studies are more fit for those that need to comply with them.

Items 3 and 4 are closely related to patient-centricity. Even though the industry is more and more frequently using that term, this concept is not yet fully applied to study protocols. But how can we put the patient in the center of a protocol—while not compromising the scientific validity? And why should we do that?

In a study conducted by Tufts CSDD [1], 15 midsized and large biopharmaceutical companies provided data on 116 Phase II and III protocols targeting diseases across multiple therapeutic areas. Medidata Solutions provided direct procedure cost data to supplement those provided by participating companies.

In total, 25,103 procedures were reviewed, and 22.3% were "non-core," i.e., a "nice to have" but not needed to address the study objectives. Tufts CSDD found that sponsors were spending US$4 to $6billion annually in direct costs to administer such "noncore" procedures that were not tied to primary or key secondary end points and regulatory requirements [2].

It should thus be in the economic interest of sponsors to reduce "waste" in the protocol. Not only to limit direct costs, but also to limit costs indirectly induced through a vicious circle triggered by unnecessary high site- and patient-burden for redundant study activities (see Fig. 11.1).

Innovation in Clinical Trial Methodologies. https://doi.org/10.1016/B978-0-12-824490-6.00001-3

FIGURE 11.1

Over-complex protocols not only produce tremendous direct costs for data collection, monitoring, processing, etc. It also triggers an increase in sample size and thus further inflates costs.

*Health professionals are less likely to refer patients to, and patients are less likely to participate in more complex clinical trials [3]. Patients are significantly less likely to sign the informed consent form when facing a more demanding protocol design [4].

Moreover, there is an inverse relationship between study design and study quality. Friedman and colleagues found, for example, that the high volume of data nowadays collected in clinical trials distracts research scientists, compromises the data analysis process, and ultimately harms data quality [5]. Nahm, Pieper, and Cunningham found that as more data are collected during protocol administration, error rates increase [6].

The solution

Patient-centricity has various components:

The informed consent process should move away from pure text layout but utilize illustrations, ideally also movies. Nowadays, the provision of information can also be delivered to the patient's home through electronic consent webpages and smartphones (see Fig. 11.2). Such use of technology also minimizes the need for travel to the site for the simple sake of getting initial study-related information.

Site visits are the most relevant burden for any study subject. Patients still at work often need to take days off only for sitting in a hospital waiting room. Elderly patients are not hot on day-long travel in busy trains or airplanes, even

FIGURE 11.2

The informed consent process should make use of patient-friendly technology, available at any time at the patient's fingertips.

less in pandemic situations. Even though weekly visits as in below real-life example may provide a nice set of data points, it is not realistic to expect patients spending at least 13 days on site (with additional time for travel) over a period of a quarter (Fig. 11.3).

The solution is to conduct all eligible visits at the patient's home (also see Melissa Hawking's section "Home nursing replacing site visits"). But some assessments such as imaging, complex pulmonary function tests, detailed physical exams, etc., require high-tech equipment or a physician who cannot easily be "brought to the patient." Some of these limitations can be handled through telemedicine (see section "Telemedicine replacing site visits").

An alternative approach is to challenge the necessity of each of these complex assessments for each scheduled visit. Some of these activities only doable on site can be replaced through other mobile technologies. Most of these, patients can easily apply themselves after a short instruction. As examples, motor function test can get replaced by actimetric wearables, on-site pulmonary function tests may get replaced by nocturnal oxygenation measure.

Reduce the number of assessments per each visit day:

A first step is the decision what assessments are "core." This selection process should be as evidence-based as possible. That means a grading is done on what assessment is directly supporting the primary endpoint and which of these add unique value—while others may only provide similar and overlapping information. This

Screening	Baseline	Treatment Period										
V 1	**V 2**	**V 3**	**V 4**	**V 5**	**V 6**	**V 7ᵃ**	**V 8**	**V 9ᵃ**	**V 10**	**V 11ᵃ**	**V 12**	**V 13ᵃ**
		Wk 1	Wk 2	Wk 3	Wk 4	Wk 5	Wk 6	Wk 7	Wk 8	Wk 9	Wk 10	Wk 11
-28 to -1	D 1	D 8	D 15	D 22	D 29	D 36	D 43	D 50	D 57	D 64	D 71	D 78
		+/-2d	+/-2d	+/-2d	+/-2d	+/-2d	+/-2d	+/-2d	+/-2d	+/-2d	+/-2d	+/-2d
X												
X												
X	X											
X												
	X											
X												
	X	X	X	X	X							
	X	X	X	X	X	X	X	X	X	X	X	X
							X		X		X	
				X			X		X		X	
X	X	X	X	X	X	X	X	X	X	X	X	X
X	X	X										
X	X	X	X	X	X	X	X	X	X	X	X	X
X	X	X	X	X	X	X	X	X	X	X	X	X
X	X	X	X	X	X		X		X		X	
X	X		X		X				X			
X	X											
X	X				X				X			
					X				X			
					X				X			
	X											
X	X				X				X			
X	X	X	X	X	X		X		X		X	
X												
X	X	X	X	X	X		X		X		X	
X												

FIGURE 11.3

Very intense visit schedules like this exclude a high number of otherwise medically eligible patients and create an unwanted selection bias.

also depends on the stage of development: quality of life scales which are relevant for health technology assessment are of minor relevance in proof-of-concept studies.

The next step is to define how frequent each assessment at minimum has to happen throughout the course of the trial. As common practice, three data points (baseline, midpoint, end of treatment) should be adequate. It is not appropriate to "fill up" visits which may be required more frequently for safety reasons with additional efficacy assessments, only because "the patient is anyway on site." Such safety visits typically only including blood draw and vital signs are ideal for a home nurse to perform—unless being overloaded with redundant other activities.

Ideally, the thus optimized visit and assessment schedule will undergo a pressure test, involving patients.

In a "Test Run" the volunteering patients will not undergo any invasive assessment, but the procedure will be done "as if." That gives a quite precise estimate of how long each measurement may take. Adding up these numbers should allow all activities within 4 h, since wait times in between the activities are coming on

Per Patient Visit Costing				Day Window	-12 to -7	-7 to -1	1	14 ±3d	28 ±3d	56 ±3d	84 ±3d
Procedures	Nurse Time (mins)	Physician Time (mins)	Data/Admin Time (mins)	Total Calculated time per activity (mins)	V1 Screen	Baseline Phase	V2 Rand Visit	Visit 3	Visit 4 (TC)	Visit 5 (TC)	Visit 6
Gen/Safety											
Informed consent	0	30	5	35	x						
Demographics	0	5	5	10	x						
Medical history	0	15	5	20	x						
Previous meds	0	10	5	15	x						
Con meds	0	10	5	15	x		x	x	x	x	x
Physical exam	0	30	5	35	x		x				x

	V1 Screen	Baseline Phase	V2 Rand Visit	Visit 3	Visit 4 (TC)	Visit 5 (TC)	Visit 6
Total Time Per Subject Per Visit (Mins):	390	40	405	70	70	70	335
Total Time Per Subject Per Visit (Hours):	6.50	0.67	6.75	1.17	1.17	1.17	5.58
Total Nurse Time Per Subject Per Visit (Mins):	60	10	60	10	10	10	20
Total Dr Time Per Subject Per Visit (Mins):	230	20	245	40	40	40	235
Total DC/Admin Time Per Subject Per Visit (Mins):	100	10	100	20	20	20	80

FIGURE 11.4

A draft protocol test-run involving real patients allows a detailed timing how long a visit may take. In that example, only the 6 assessments of a more complex protocol are shown. The sum adds up to a total beyond 4 hrs what creates a risk not all activities can get completed during a single day.

top. Otherwise, a single day will not suffice to complete all activities with appropriate quality (Fig. 11.4).

Entry and exit patient interviews: Entry patient interview is a short structured interview about why the patient has interest to participate in the given study. Most easily, this can happen before and after a dry run. The exit interview covers items like "What did you like about that day?" and "How could we make this visit more attractive for you?"

SWOT analysis

Strengths: Better-written protocols will enhance compliance with the requirements and improve enrollment rates and data quality.

Weaknesses: It will take more time to create a patient-centric protocol and also frustrate some mainly academically interested contributors, since their questions may not get answered.

Opportunities: Make studies cheaper and more "fit for purpose," thus also increasing the probability of a positive outcome.

Threats: In case competent authorities request additional supportive data after first review, the clinical database may not hold such additional data in a streamlined and slimmed-down design that collects less (initially not needed) data points. Upfront scientific advice will massively reduce that risk.

Take-home message

In short: LESS is MORE.

Most existing study protocols fulfill the ICH-GCP requirements, but they tend to be over-complex. Imprecise and overloaded protocols are a main risk for slow enrollment and data variability, leading to a failed study. The attempt to streamline the development process should start with better planning, i.e., the implementation of a defined protocol development process that deserves that name. That shall include a rigid evaluation of the required visits and assessments—and which of these actually have to happen on site or can also be done at patient's home. Input from patients makes this evaluation more reliable. That would also allow for less site monitoring efforts, which will contribute to major cost savings.

References

[1] Getz K, Stergiopoulos S, Marlborough M, Whitehall J, Curran M, Kaitin K. Quantifying the magnitude and cost of collecting extraneous protocol data. Am J Ther 2015;22(2): 117−24.

[2] Getz KA, Kaitin KI. The impact of bad protocols. In: Schueler P, Buckley B, editors. Re-engineering clinical trials. Elsevier; 2015. p. 105−15.

[3] Ross S, Grant A, Counsell C, Gillespie W, Russell I, Prescott R. Barriers to participation in randomized controlled trials -a systematic review. Clin Epidemiol December 1999; 52(12):1143−56.

[4] Madsen S, Holm S, Riis P. Ethical aspects of clinical trials. Attitudes of public and out-patients. J Intern Med June 1999;245(6):571−9.

[5] Friedman L, Furberg C, DeMets D. Data collection and quality control in the fundamentals of clinical trials. Chapter 11. Springer Science and Business Media; 2010. p. 199−214.

[6] Nahm ML, Pieper CF, Cunningham MM. Quantifying data quality for clinical trials using electronic data capture. PLoS One 2008;3(8):e3049. https://doi.org/10.1371/journal.pone.0003049.

The use of new digital endpoints

12

Bill Byrom, BSc, PhD

Vice President of Product Strategy and Innovation, Signant Health, London, United Kingdom

The need

We strive to gain pertinent insights into drug development to make appropriate investment decisions in a timely manner, and to collect data that can effectively support regulatory decision making and drug labeling. Failure rates in clinical drug development remain high—estimated by Hay et al. to be 89.6%, with a 66.6% attrition rate between phase II and III [1]. In an analysis of new drug applications approved by the FDA between 1985 and 1997, Khan et al. [2] reported that over 50% of adequate-dose new antidepressant treatment arms failed to demonstrate statistically significant separation from placebo. New endpoints and richer data, with greater sensitivity to detect change and less influenced by placebo response, may help to understand intervention effects more completely and mitigate these failure rates, and provide sufficient information to enable development programs to fail fast when appropriate and save unnecessary patient exposure and costs.

By leveraging new technologies to develop new endpoint measures, as discussed in this chapter, we may have the possibility to measure things with greater sensitivity than before, or to measure things that were previously hard or impossible to measure. We may also have the possibility to leverage these technologies outside the clinic environment, within the patients' homes, for example. This affords the opportunity to measure aspects of health status and functioning more frequently than during routine clinic visits—potentially providing a richer picture of treatment effects—and to study elected patient functioning in real-world conditions as a complementary measure to established in-clinic tests. Measuring in the home environment may also have additional measurement advantages—such as eliminating the "white coat effect" as reported for blood pressure measurement in hypertension patients [3], improving patient monitoring, and making trial participation more convenient for patients by potentially reducing the frequency of in-person clinic visits.

The collection of frequent or continuous measurements from trial participants using new technologies may change how clinical trials are designed and conducted, and may change the endpoints we develop to measure the concepts of interest associated with these trials.

Innovation in Clinical Trial Methodologies. https://doi.org/10.1016/B978-0-12-824490-6.00007-4
91

The COVID pandemic accelerated the wish for new and more robust trial concepts which are less dependent on patient travel, thus making data collection less vulnerable and at the same time more patient-friendly.

The solution

We live in a highly connected world. Many of the things we interact with on a routine basis contain electronic sensors that are collecting data on a regular or continuous basis. Our smartphones, for example, contain multiple sensors which, while they have a purpose in enabling certain hardware and software capabilities, can also be leveraged for other purposes. Freeman Dyson, a leading contemporary theoretical physicist and mathematician, famously stated, "The year 2000 was essentially the point at which it became cheaper to collect information than to understand it" [4]. While stated in the context of genetic information, the parallels can be seen in the amount of data and information that new technologies, such as those containing sensors, can collect and report. The challenge is in understanding it, or in our context, determining the clinical value, interpretation and relevance of the data collected and its associated derived endpoints.

In this section we review a number of new technologies and their potential for the development of novel, sensitive, and informative endpoints for clinical research.

Wearables and remote sensors

The miniaturization of sensors and circuitry has led to the rise in availability and utilization of connected sensors over the last decade. While this has been seen across many industries, we see its relevance to clinical research in the rapidly expanding health and wellness market that produces wearables and other sensors aimed at measuring activity, heart rate, sleep, and other aspects of personal health. The global connected health and wellness devices market was estimated to be $123.2 billion in 2015 and is expected to reach $612.0 billion by 2024 [5].

The pharmaceutical industry has been slower to adopt these sensor-based technologies at scale. This may be in part due to concerns over demonstrating the validity of outcomes data collected using wearable and remote devices to satisfy regulatory examination; and in part due to concerns over site or patient burden or complexity. Despite this, wearables and remote sensors offer great potential to gain deeper insights into patient functioning for both in-clinic and at-home assessments.

In-clinic functional performance test instrumentation

There are a number of functional performance tests often conducted during clinic visits within clinical trials aimed at measuring aspects of mobility and movement— such as the 6-min walking test (6MWT) and the timed up-and-go (TUG) test.

Instrumentation of many of these tests using wearables and other sensors can provide deeper insights than originally obtainable via the traditional clinician-assessed tests.

For example, the TUG has been used to screen for gait and balance issues in older adults [6] and to assess the risk of falls in patients with Parkinson disease [7]. This test requires the patient to complete the following procedure as quickly and safely as possible: get up from a chair (with armrests), walk 3 m, turn 180 degrees, walk back to the chair, and sit back down. The time to complete the test is the main outcome measure and is typically recorded by a clinician using a stopwatch. However, TUG time has been shown in other studies to have only limited ability to assess fall risk in some populations—for example, in community-dwelling older adults [8]. Overall time to complete the test may be a blunt summary and does not provide insights into specific aspects of mobility that may be exposed by conduct of the test—such as measures of balance and the number of steps taken to conduct the 180 degrees turn. Instrumenting the test using sensors can provide these deeper insights and enhance the ability to predict outcomes such as falls [9]. For example, Greene et al. [10] used sensor units containing both an accelerometer and a gyroscope attached to each leg, below the knee, to instrument the TUG test in patients with Parkinson's disease. By comparing to a 6-month fall diary, they were able to demonstrate 73% accuracy in predicting falls within 90 days of the baseline assessment by deriving estimates of falls risk and frailty using the enhanced insights provided by the sensor data. Some of the additional measures that can be derived from this instrumented TUG test are illustrated in Fig. 12.1.

At-home monitoring

Wearables and mobile sensors provide the opportunity to measure clinical endpoints outside the clinic environment. This can enable measurements to be made in free-living conditions as opposed to performance testing in clinic. For example, understanding the activity patterns that patients elect to conduct during their daily lives could be measured by a wearable accelerometer as an alternative to, or to complement, an in-clinic functional performance test such as a treadmill test or 6MWT using a corridor circuit. In this example it may be considered that, in some cases, free-living activity measures may be more pertinent and informative than functional capacity testing, and may be closely related to quality of life and activities of daily living. At-home testing also affords the possibility to measure constructs more frequently which may provide a richer picture of intervention effects—for example, daily measurement of blood pressure and heart rate as opposed to measurement every few weeks during a clinic visit. Finally, wearables and sensors may provide the opportunity to measure things that were not possible to measure before. For example, a continuous glucose monitor sensor can provide rich information on glycemic control over the day, which may be more informative than average measures derived from HbA1c laboratory values.

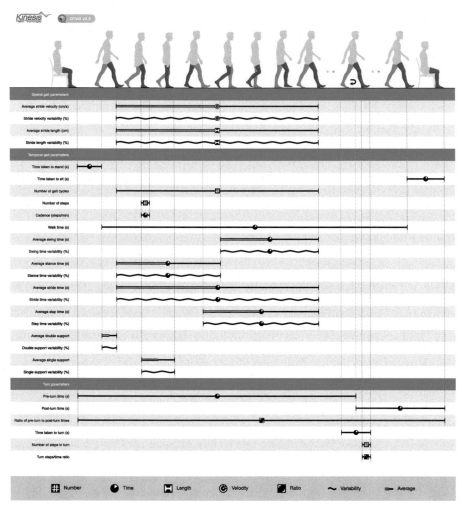

FIGURE 12.1

Chart illustrating some of the additional metrics that can be derived from an instrumented timed up-and-go test using a sensor attached to each leg below the knee.

Image reproduced with permission of Kinesis Health Technologies Ltd, Dublin, Ireland.

Good progress is being made on defining evidentiary requirements to support the selection and implementation of wearables and remote sensors to generate clinical endpoints for use in regulatory decision making by the Critical Path Institute's Electronic Patient-Reported Outcomes (ePRO) Consortium [11], the Clinical Trials Transformation Initiative [12], and the Drug Information Association's Study Endpoints Community [13].

Activity and sleep measurement are worthy of further expansion in this section as these measures have importance across a broad range of disease indications.

Measuring activity and sedentary behavior

Accelerometers provide the most commonly used means to assess sedentary behavior and activity patterns in patients in free-living conditions. The choice of device and its usage is, in part, dependent upon the measurement concept of interest. For example, if studying sedentary behavior, some advantages can be seen by selecting a device that can be conveniently attached to the thigh as this enables an accelerometer to also act as an inclinometer and distinguish sitting and lying from standing. The ActivPAL device (PAL Technologies, Glasgow, UK), for example, is worn under a Tegaderm dressing providing permanent waterproof attachment for a number of days without removal [14]. A second example—the amount of wear time required to provide robust measures of activity—also depends on the concept of interest for measurement. If our interest is overall daily activity counts or steps, or daily time in different levels of exercise intensity, then ensuring the device is worn for the majority of the awake interval is important. If, however, our concept of interest is real-world walking speed, or average walking cadence (stepping rate), then wearing the device during a number of bouts of purposeful walking per day may suffice. By improving accuracy of mobility assessment with accelerometer devices we can achieve a magnitude of insight information about long-term mobility in different populations, also including frail and elderly patients. Keppler et al. [15], for example, reported the acceptability of algorithms using accelerometer data to detect steps in an orthogeriatric population with a median age of 75 years.

It is important to remember that additional validation considerations may need to be examined when using an accelerometer in certain populations. For example, if gait patterns differ significantly from populations for which validation data exist (e.g., the shuffling gait commonly observed in some Parkinson's disease patients), additional validation evidence may be important to demonstrate that steps and stepping behavior can be accurately detected. In addition, as with any clinical endpoint, important properties such as interpretability (e.g., meaningful within-patient change) will differ between patient populations and require additional estimation where not already understood.

Promising areas of innovation include the embedding of pressure sensors and accelerometers into footwear insoles such as the F-Scan system (Tekscan Inc., South Boston, MA) and Moticon's insole (Moticon GmbH, Munich, Germany). These approaches may provide richer gait information than is possible using a wrist or waist-worn accelerometer and may facilitate data collection in a frictionless manner in free-living settings. Studies using the Moticon sensor-instrumented insole solution, for example, have demonstrated good validation and reliability of gait parameters collected [16,17].

Measuring sleep

Sleep architecture refers to the basic structural organization of normal sleep, consisting of alternating periods of rapid eye movement (REM) and non-REM (NREM) sleep [18]. In addition to sleep architecture, sleep quality, sleep quantity, circadian rhythmicity, sleep consolidation, regularity, and napping are also important factors in assessing sleep and wake patterns. A number of outcome measures are commonly estimated to assess sleep quality and quantity including sleep onset latency, wake after sleep onset, sleep efficiency, number of awakenings, and total sleep time. Wrist-worn accelerometers have shown promise in estimation of sleep quality, quantity, and circadian rhythms in the home setting [19], but not for assessing sleep architecture (REM and NREM Sleep). Actigraphy-determined sleep parameters correlate well with polysomnography data in normal adults [20], but less so for other populations such as insomniacs where the correlation can be as low as 50% for some of the parameters such as sleep onset latency [19,21]. Actigraphy-based sleep estimation depends upon detection of periods of movement and immobility, where periods of movement help to identify periods of wakefulness (Fig. 12.2). While

FIGURE 12.2

An actogram showing detection of sleep periods over a seven-day interval for a single individual. Orange bars represent periods of activity. Shaded blue areas represent resting/sleep periods.

Image reproduced with permission of ActiGraph LLC, Pensacola, FL, USA.

actigraphy generally provides good prediction of sleep periods, it is less sensitive to distinguishing between sleep and "still wakefulness," and as a result can overestimate total sleep time [18]. Despite limitations, actigraphy provides a valuable and practical approach to determination of objective sleep parameters in free-living settings and large populations.

A new class of sleep assessment tools have recently emerged that claim to measure sleep parameters without the requirement to be worn. For example, the Beddit 3 Sleep monitoring system (Apple, Cupertino, CA) uses a flexible piezoelectric film sensor that is placed beneath the bed sheet which measures the forces caused by the body on the bed to detect tiny movements that can be interpreted to estimate pulse (heart pumping), breathing effort (thorax extension), and overall body movement. Validation work comparing heart rate estimates to ECG [22] and breathing effort to the respiration effort signal in PSG [23] show some promise but more work is needed.

A second approach used by monitors such as the S+ sleep sensor (ResMed, San Diego, CA), the EZ-Sleep [24] and Emerald AI (Emerald Innovations, Boston, MA, USA) analyses the reflection of transmitted wireless radiofrequency waves to monitor the movements of the patient in bed such as the expansion and relaxation of the chest during respiration, and overall body movements such as positional changes, arm twitches, and shrugs to determine sleep parameters. Early studies using EZ-Sleep have shown good accuracy in the estimation of sleep latency and total sleep time [24]. The same technology approach has been used to measure at-home movement and mobility in care home settings [25].

Smartphone sensors and performance outcomes

Modern smartphones contain multiple inbuilt sensors to enhance user experience and provide specific smartphone features (Fig. 12.3). For example, tri-axial accelerometers are used to determine the spatial positioning of the smartphone to enable the screen display to switch between portrait and landscape orientation when the handset is rotated. More recently, novel application of these inbuilt sensors has enabled new and inventive uses for the smartphone in the area of health and wellness. For example, the same accelerometer sensor can be accessed by apps and the sensor output used to detect walking activity and estimate the number of steps taken. This provides an exciting opportunity to develop new ways of measuring health outcomes and derive new clinical endpoints in clinical trials.

Apple (Apple Inc., Cupertino, CA, USA) and Google (Google, Mountain View, CA, USA) have made it easy to access smartphone and tablet sensor data to create health apps. A number of high profile studies leveraging Apple Research Kit, for example, have been helpful in illustrating the potential to develop performance tests that can be conducted using a smartphone alongside other measures such as patient-reported outcomes. The MyHeartCounts study (Stanford Medicine, Stanford, CA, USA), for example, leveraged GPS and accelerometer sensors to deliver a "six minute walking test" designed to be conducted out of doors and independently [26].

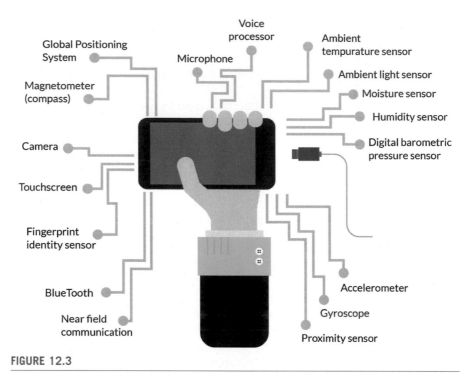

FIGURE 12.3

Typical sensors contained in modern smartphones.

A second compelling example is the application to study Parkinson's disease (PD) developed by Roche (F. Hoffmann-La Roche Ltd., Basel, Switzerland) in collaboration with academic research groups [27]. This Android app leverages a number of sensors and components in the patient's own smartphone to measure aspects of health status and symptomology while conducting a number of short performance tasks. Tasks include a phonation test to measure voice degeneration, simple tests of balance and gait using the accelerometer to measure sway and stepping, a finger tapping dexterity test using the smartphone touchscreen, and tests to measure tremor using the device accelerometer while the patient holds their smartphone with arm extended for 30 s. Smartphone technology provides a convenient approach to implement, collect, and transmit sensor data for each task, but additional complex work is required to develop and validate algorithms to interpret the data collected and translate the data into clinical outcome measures.

Frequent objective measurement is valuable in the assessment of PD where symptoms important in optimizing treatment may not be observed every day or during routine clinic assessments. Smartphone-derived performance outcome measures afford an opportunity to develop a range of objective measures using the smartphone sensors and touchscreen. Because a growing number of clinical trials utilize smartphone technology to collect patient-reported outcome measures using dedicated apps, incorporating performance tests using the same platform presents a convenient approach to collect additional objective measures in combination with patient self-reports.

Video gaming platforms

Motion-based gaming platforms use depth-cameras to detect body movements and enable users to interact with gaming applications in more immersive ways. The same depth-camera technology, and their associated software development kits (SDKs), can be used to develop custom software with application in health (Fig. 12.4). To date, the most commonly used solution is the Microsoft Kinect depth-camera associated with the Xbox gaming system (Microsoft Corp., Redmond, Washington, USA), although reduced ongoing development by the manufacturer may lead to greater use of other more advanced (yet similarly low-cost) technologies, such as the Intel RealSense camera range (Intel Corp., Santa Clara, CA, USA) [30]. Particular utility for applications using this technology is in the field of rehabilitation where interactive games can provide a more engaging approach to continued conduct of exercising, and algorithms can be developed to determine whether exercises are being conducted correctly for optimal results. This approach has been reported to produce good outcomes in terms of reduced readmission rates in orthopedic and stroke patients [31].

A second application area of video game applications is in the measurement of cognitive function. Project:Evo, for example, is a game application developed by Akili Interactive Labs (Boston, MA, USA) that can be used to measure interference

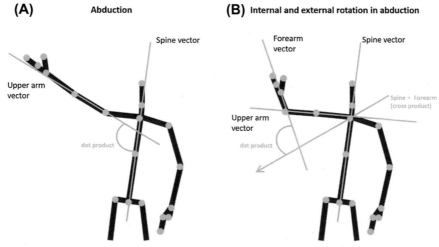

FIGURE 12.4

Estimation of shoulder ranges of motion using detection of 3D body joint positions from a gaming platform depth camera [32].

processing, a key component of executive function. It is intended to be an engaging alternative to conventional cognitive testing batteries and is currently being tested in a variety of clinical studies in multiple patient populations including ADHD, autism, depression, and traumatic brain injury.

Insights from video and voice acoustics

Speech patterns, and their changes over time, provide potential insights into the health status of patients. For example, a study of patients with PD showed that around 75% of patients exhibit some form of vocal impairment [33]; and voice acoustical analysis of voice samples from patients with extremely early-stage PD suggested that voice acoustical changes can be good predictors of early onset of the disease [34]. In patients with depression, certain aspects of speech, such as speaking rate and pitch variability, have been shown to correlate well with conventional measures of the severity of depression such as the Hamilton Depression Rating Scale [35]. One patient-reported outcome measure developed for depression self-assessment, MERET (Memory Enhanced Retrospective Evaluation of Treatment), uses a recording of how a patient feels at baseline using their own words and voice. This is played back at subsequent time points at which the patient is asked to rate their perceived change from baseline condition. Enabling patients to hear their recorded description provides additional voice acoustical cue—such as tone, hesitation, and speaking rate—that enriches their understanding of baseline state [36].

Smartphone technology has simplified the collection of digital voice samples. For example, phonation tests for PD patients have been developed in clinical research mobile apps using both Apple Research Kit [37] and on the Android platform [38]. This opens the possibility of using such inexpensive techniques in large-scale clinical trials.

More recently, machine learning techniques afford greater opportunity to derive insights from complex data such as voice samples. One study, for example, used machine learning techniques to create a model based on an initial input of 370 extracted linguistic features that was able to adequately distinguish between Alzheimer's disease patients and healthy controls based on analysis of short narrative samples elicited with a picture description task [39].

Similarly, video analysis may provide valuable health status insights. Promising work has been conducted on the extraction of facial expression based on computer recognition of the relative position of facial landmarks—for example, in aiding the diagnosis of autism spectrum disorders in young children while watching short video content designed to elicit certain emotional responses such as surprise and happiness [40]. Artificial intelligence techniques using video analysis have also been successful in measuring medication-taking behavior in which algorithms can identify medication-taking behavior by pill/capsule size and color detection and the capture of swallowing action through selfie-camera video [41].

Considerations for clinical endpoint development

It's important that the approach used to measure a clinical endpoint is determined not by a drive to use a new technology, but by the drive to seek a feasible, valid and appropriate way to measure the clinical endpoint. In common with other clinical endpoints, an endpoint's conceptual framework should identify how the endpoint is able to measure a concept of interest defined by the study objectives and how this concept of interest relates to aspects of health that are meaningful to patients [42]. Developing this framework to define pertinent and meaningful clinical endpoints then enables the final step of determination of a suitable measurement approach. In some instances, the most suitable measurement approach may be achieved using a novel technology such as a wearable or a video game, but in other cases it may be better accomplished using an in-clinic test or a patient-reported outcome measure, for example. When the intended measurement approach is new, early discussion with the relevant regulatory bodies is encouraged.

Once a novel technology-generated clinical endpoint has been selected, it is important to consider section of a technology that is fit-for-purpose, to ensure endpoint properties are well characterized, and (where this has not already been done) to compile evidence to support the use of the measurement approach and specific endpoint for regulatory decision making.

Consumer devices are often associated with advanced form factors that may be associated with increased patient acceptability. Whether a device is a consumer device, is specific to clinical research, and/or has a market certification/clearance (such as CE marking or a 510(k)) should not be a driving factor around device selection. Instead, the determination of whether a sensor, wearable, or other technology is fit-for-purpose to measure a clinical endpoint should be assessed with reference to a number of criteria. The ePRO consortium has put forward a framework that is a helpful guide [11] and their key considerations are summarized in Table 12.1.

Bring-your-own-wearable (BYOW)

In the area of patient-reported outcomes, patients' own smartphones have increasingly been used to run apps that administer patient-reported outcome instruments (BYOD—bring-your-own-device). The key consideration in BYOD is whether the measurement properties of the instruments can be assumed to be equivalent across the different screen sizes, resolutions, and makes and models of the smartphones used in the study. Increasingly, patients are using their own wearable and sensor technology for personal wellness and fitness tracking. Might it be possible to consider a similar approach in future whereby patients are able to use their own wearables in a clinical trial (BYOW—bring-your-own-wearable), and only those without one are provided one for use in the study? This may reduce costs, but would data quality and integrity be compromised?

The key consideration with BYOW is the same as BYOD—can we demonstrate that measurement properties are sufficiently aligned to enable the use of different

Table 12.1 Criteria important in determining a fit-for-purpose technology, sensor, or wearable device.

Factor	Description
Safety	
Evidence that technology is safe to use within context of use defined by the protocol	Manufacturer information including, as applicable, mechanical, electrical, and biological engineering performance, such as fatigue, wear, tensile strength, and compression; electrical safety and electromagnetic compatibility; sterility; and stability/shelf-life.
Suitability	
Study design factors	• Battery length/storage capacity meets needs of intended use. • Setup and maintenance processes are suitable for sites/patients. • Data acquisition: is real-time data/remote access to data required for patient monitoring? • Data blinding: should patients be able to see the data collected by the technology, or should this be blinded to the patient?
Usability and feasibility	• Form factor, usage/wear location, period of time required to use the sensor is acceptable to the intended patient population. • Ease of use and maintenance including operation, wear, charging, data transmission. • Suitability of training procedures and instructions. • Feasibility in the context of the clinical trial protocol—e.g., potential burden on patients and sites.
Sensor vendor factors	• Firmware/software version control—can technology be maintained on a defined firmware version for the duration of the study and, if not, might this influence the calculation of clinical outcomes generated? • Compliance with 21 CRF Part 11 and other relevant regulations relating to data security, traceability, and data protection. • Deployment/logistics support. • Vendor risk assessment—financial viability/ability to continue to access sensor data for duration of study. • Cost.
Clinical endpoint factors	The technology measures outcomes suitable to derive the clinical endpoint defined in the protocol (e.g., supports specific wear location to detect sitting and standing).

Table 12.1 Criteria important in determining a fit-for-purpose technology, sensor, or wearable device.—*cont'd*

Factor	Description
Analytical validity of sensor output	
Accuracy	• Outcome measures derived are sufficiently accurate with reference to a standard (e.g., concurrent validity evidence). • In some cases, specific accuracy evaluation in the patient population may be needed—e.g., to ensure algorithms can accurately detect steps in patient groups that exhibit different gait patterns (e.g., shuffling gait in Parkinson's disease patients).
Precision	Outcome measures derived have adequate reproducibility.
Reliability	Satisfactory inter- and intra-unit reliability. Manufacturing processes follow quality standards to ensure ongoing reliability.

devices within the same study to measure the endpoint of interest. For example, if measuring activity or sleep, could patients with their own Fitbit (Fitbit Inc., San Francisco, CA, USA), Garmin (Garmin International Inc., Olathe, KS, USA), or Apple watch (Apple Inc., Cupertino, CA, USA) in place of a study provisioned device?

If we consider activity measurement using wearable accelerometers as an example, there are a number of considerations. If all devices provide access to raw (unprocessed) sensor data, is there sufficient evidence that the raw data generated is sufficiently equivalent across devices, and is the application of common processing algorithms on the raw data then sufficient to ensure comparability across devices? As many devices use the same internal accelerometer sensors, this seems plausible perhaps.

Where devices only provide processed data, which is more common with consumer devices that typically use undisclosed proprietary algorithms to provide measures derived from the raw sensor data—is there sufficient evidence that the outcome measures provided are sufficiently equivalent to those provided by other devices? Currently, it would appear that step counts collected using different devices can vary significantly between makes and models of devices. For example, Bender et al. [43] studied concordance to outcomes measures provided by Fitbit Flex, Fitbit Charge HR, Garmin vivoactive, and Apple Watch in healthy volunteers in free-living conditions for 14 days. They concluded that step count, distance traveled, and calories burned could vary significantly between devices used concurrently. While within-patient change is typically of interest, interpreting the magnitude of within-patient change observed would be difficult if devices were measuring with different degrees of bias and without a way of standardizing measurements to a common scale. At this point, we are likely some way from a BYOW approach, at least for endpoint measurement to support regulatory decision making.

Endpoint property considerations

Ensuring that the properties of clinical endpoints derived from novel technologies are well understood in reference to the target population is essential for their use in clinical development to support new drug applications and regulatory decision making. Such properties include:

Construct validity. Demonstrating that the clinical endpoint is able to measure the concept of interest as defined in the study protocol and the endpoint conceptual framework. This may include, for example, comparison to another recognized measurement method (concurrent validity). In some cases, this evidence may be population-specific. For example, the ability to detect the incidence and timing of steps during walking activity using an accelerometer may need additional evaluation if used in PD patients where walking patterns often comprise a shuffling action as opposed to a more conventional walking movement.

Ability to detect change. Clinical endpoints should be sensitive to detect change when change exists. This can typically be demonstrated through studies of an intervention known to create a change in the concept of interest.

Clinical interpretation. Understanding the magnitude of within-patient change that is considered meaningful to patients is vital to the interpretation of endpoint changes observed in clinical studies. Meaningful within-patient change will be population dependent. For example, an increase of 500 steps per day may be meaningful to a patient with COPD but not meaningful to someone with migraine.

SWOT analysis

Strengths	Weaknesses
• Generate objective measures in the home setting. • Measure more frequently. • Generate richer insights than via conventional methods. • In some cases measure passively—with little impact on the patient.	• Increased site and/or patient burden. • Added expense. • Increased study complexity. • Validation work may be needed to support the approach.

Opportunities	Threats
• Learn more about intervention effects earlier in clinical drug development. • Measure constructs that have been hard or difficult to measure previously. • Enhanced between-visit patient monitoring. • Reduced number of clinic visits due to increased remote assessment.	• Untested regulatory acceptability.

Applicable regulations

1. Food and Drug Administration. Guidance for Industry: Computerized Systems Used in Clinical Investigations. www.fda.gov/OHRMS/DOCKETS/98fr/04d-0440-gdl0002.pdf
2. Food and Drug Administration. Guidance for Industry: Electronic Source Documentation in Clinical Investigations. www.fda.gov/downloads/drugs/guidancecompliancyregulatoryinformation/guidances/ucm328691.pdf
3. Food and Drug Administration. 21 CFR Part 11. https://www.accessdata.fda.gov/scripts/cdrh/cfdocs/cfcfr/CFRSearch.cfm?CFRPart=11
4. Food and Drug Administration. Guidance for Industry Part 11, Electronic Records; Electronic Signatures — Scope and Application. https://www.fda.gov/media/75414/download
5. Food and Drug Administration. Guidance for Industry: Patient-Focused Drug Development: Collecting Comprehensive and Representative Input. https://www.fda.gov/media/139088/download

Take—home message

• Our current regulatory framework identifies the steps needed to develop and validate clinical endpoints. These approaches can be used to ensure that new digital endpoints derived using novel technologies can be implemented appropriately, and with sufficient supporting evidence, in clinical development programs.
• New digital endpoints from novel technologies provide opportunity to (i) measure constructs that cannot be measured feasibly using existing methods, (ii) measure objectively and with the potential of greater precision compared to some traditional methods, and (iii) measure more frequently and provide additional

richness to the insights that can be gained compared to traditional measurement approaches. New technologies may also facilitate greater remote measurement which may enhance patient monitoring and make study participation more convenient.

- New technologies should be used to measure clinical endpoints not because it would be trendy or interesting to do so, but when they offer a viable approach to appropriately measure the endpoints of interest.
- Sponsors should select a "fit-for-purpose" technology with respect to the construct being studied, the protocol, and the study population, and use this evaluation to support their approach.
- Ensuring that the properties of clinical endpoints derived from novel technologies are well understood in reference to the target population is essential for their use in clinical development to support labeling claims and regulatory decision making.
- We understand enough to begin using these approaches where appropriate and to begin gaining acceptance for the clinical endpoints they can measure.

References

[1] Thomas DW, Hay M, Craighead JL, et al. Clinical development success rates for investigational drugs. Nature Biotechnology 2014;32(1):40−51.

[2] Khan A, Leventhal RM, Khan SR, et al. Severity of depression and response to antidepressants and placebo: an analysis of the Food and Drug Administration database. J Clin Psychopharmacol 2002;22(1):40−5.

[3] Schwartz JE, Burg MM, Shimbo D, et al. Clinic blood pressure underestimates ambulatory blood pressure in an untreated employer-based US population: results from the masked hypertension study. Circulation 2016;134(23):1794−807.

[4] Byrom B. Leveraging technology to develop new trial endpoints. Appl Clin Trials 2018; 27(12):28−31.

[5] Grand View Research. Connected health and wellness devices market report. August 2016.

[6] Podsiadlo D, Richardson S. The timed "Up & Go": a test of basic functional mobility for frail elderly persons. J Am Geriatr Soc 1991;39:142−8.

[7] Nocera JR, Stegemöller EL, Malaty IA, et al. Using the timed up and go test in a clinical setting to predict falling in Parkinson's disease. Arch Phys Med Rehabil 2013;94: 1300−5.

[8] Barry E, Galvin R, Keogh C, et al. Is the timed up and go test a useful predictor of risk of falls in community dwelling older adults: a systematic review and meta-analysis. BMC Geriatr 2014;1:14.

[9] Greene B, McManus K, Redmond SJ, et al. Digital assessment of falls risk, frailty, and mobility impairment using wearable sensors. NPJ Digital Med 2019;2:125. https://doi.org/10.1038/s41746-019-0204-z.

[10] Greene BR, Caulfield B, Lamichhane D, et al. Longitudinal assessment of falls in patients with Parkinson's disease using inertial sensors and the timed up and go test. J Rehabil Assist Technol Eng 2018;12(5). https://doi.org/10.1177/2055668317750811.

[11] Byrom B, Watson C, Doll H, et al. Selection of and evidentiary considerations for wearable devices and their measurements for use in regulatory decision making: recommendations from the ePRO consortium. Value Health 2018;21:631—9.

[12] Clinical Trials Transformation Initiative (CTTI). Recommendations on developing technology derived novel endpoints. 2017. https://www.ctti-clinicaltrials.org/briefingroom/recommendations/developing-novel-endpoints-generated-mobile-technologyuse-clinical. [Accessed 25 August 2019].

[13] Walton MK, Cappelleri JC, Byrom B, et al. Considerations for development of an evidence dossier to support the use of mobile sensor technology for clinical outcome assessments in clinical trials. Contemporary Clinical Trials 2020;91:105962.

[14] Byrom B, Stratton G, McCarthy M, et al. Objective measurement of sedentary behaviour using accelerometers. Int J Obes 2016;40(11):1809—12.

[15] Keppler AM, Schieker M, Nuritidinow T, et al. Validity of accelerometry in step detection and gait speed measurement in orthogeriatric patients. PLoSOne 2019;14(8): e0221732. https://www.ncbi.nlm.nih.gov/pmc/articles/PMC6716662/.

[16] Braun BJ, Veith NT, Hell R, et al. Validation and reliability testing of a new, fully integrated gait analysis insole. J Foot Ankle Res 2015;8:54—60.

[17] Oerbekke MS, Stukstette MJ, Schutte K, et al. Concurrent validity and reliability of wireless instrumented insoles measuring postural balance and temporal gait parameters. Gait Posture 2017;51:116—24.

[18] Byrom B, Mc Carthy M, Schueler P, et al. Brain monitoring devices in neuroscience clinical research: the potential of remote monitoring using sensors, wearables, and mobile devices. Clin Pharmacol Ther 2018;104(1):59—71.

[19] Martin JL, Hakim AD. Wrist actigraphy. Chest 2011;139:1514—27.

[20] Ancoli-Israel S, Cole R, Alessi C, et al. The role of actigraphy in the study of sleep and circadian rhythms. Sleep 2003;26:342—92.

[21] Sivertsen B, Omvik S, Havik OE, et al. A comparison of actigraphy and polysomnography in older adults treated for chronic primary insomnia. Sleep 2006;29:1353—8.

[22] Paalasmaa J, Toivonen H, Partinen M. Adaptive heartbeat modeling for beat-to-beat heart rate measurement in ballistocardiograms. IEEE J Biomed Health Inform 2015; 19:1945—52.

[23] Paalasmaa J, Leppäkorpi L, Partinen M. Quantifying respiratory variation with force sensor measurements. In: 33rd Annual International Conference of the IEEE Engineering in Medicine and Biology Society; 2011.

[24] Hsu C-Y, Ahuja A, Yue S, et al. Zero-effort in-home sleep and insomnia monitoring using radio signals. In: Proceedings of the ACM on Interactive, Mobile, Wearable and Ubiquitous Technologies; Vol. 1, No. 3; 2017. Article 59. Publication date: September 2017.

[25] Kabelac Z, Tarolli CG, Snyder C, et al. Passive monitoring at home: a pilot study in Parkinson disease. Digital Biomark 2019;3:22—30.

[26] McConnell MV, Shcherbina A, Pavlovic A, et al. Feasibility of obtaining measures of lifestyle from a smartphone app: the MyHeart counts cardiovascular health study. JAMA Cardiol 2017;2(1):67—76.

[27] Roche. Roche app measures Parkinson's disease fluctuations. www.roche.com/media/store/roche_stories/roche-stories-2015-08-10.htm.

[28] Roche. Smart biomarkers and innovative disease-modifying therapies for Parkinson's disease. www.nature.com/nature/outlook/parkinsons-disease/pdf/roche.pdf.

[29] Gravitz L. Technology: monitoring gets personal. Nature October 27, 2016;538:S8—10.

[30] Breedon P, Byrom B, Siena L, Muehlhausen W. Enhancing the measurement of clinical outcomes using Microsoft Kinect. In: International Conference on Interactive Technologies and Games (iTAG) 2016. IEEE Xplore; 2016. http://ieeexplore.ieee.org/document/7782516.

[31] Jintronix. http://www.jintronix.com/wp-content/uploads/2016/07/TNJH-Case-Study.pdf.

[32] Byrom B, Walsh D, Muehlhausen W. New approaches to measuring health outcomes — leveraging a gaming platform. J Clin Stud 2016;8(6):40—2.

[33] Ho AK, Iansek R, Marigliani C, et al. Speech impairment in a large sample of patients with Parkinson's disease. Behav Neurol 1998;11(3):131—7.

[34] Harel B, Cannizzaro M, Snyder PJ. Variability in fundamental frequency during speech in prodromal and incipient Parkinson's disease: a longitudinal case study. Brain Cognit 2004;56(1):24—9.

[35] Cannizzaro M, Harel B, Reilly N, et al. Voice acoustical measurement of the severity of major depression. Brain Cognit 2004;56(1):30—5.

[36] Mundt JC, DeBrota DJ, Greist JH. Anchoring perceptions of clinical change on accurate recollection of the past: memory enhanced retrospective evaluation of treatment (MERET). Psychiatry (Edgmont) 2007;4(3):39—45.

[37] Bot BM, Suver C, Neto EC, et al. The mPower study, Parkinson disease mobile data collected using ResearchKit. Sci Data 2016;3:160011. https://doi.org/10.1038/sdata.2016.11.

[38] Lipsmeier F, Taylor KI, Kilchenmann T, et al. Evaluation of smartphone-based testing to generate exploratory outcome measures in a phase 1 Parkinson's disease clinical trial. Mov Disord 2018;33(8):1287—97.

[39] Fraser KC, Meltzer JA, Rudzicz F. Linguistic features identify Alzheimer's disease in narrative speech. J Alzheim Dis 2016;49:407—22.

[40] Hashemi J, Campbell K, Carpenter KLH, et al. A scalable app for measuring autism risk behaviours in young children: a technical validity and feasibility study. In: MOBIHEALTH'15: Proceedings of the 5th EAI International Conference on Wireless Mobile Communication and Healthcare; December 2015. p. 23—7.

[41] Labovitz DL, Shafner L, Reyes Gil M, et al. Using artificial intelligence to reduce the risk of nonadherence in patients on anticoagulation therapy. Stroke 2017;48(5):1416—9.

[42] Walton MK, Powers JH, Hobart J, et al. Clinical outcome assessments: conceptual foundation—report of the ISPOR clinical outcomes assessment — emerging good practices for outcomes research task force. Value Health 2015;18:741—52.

[43] Bender CG, Hoffstot JC, Combs BT, et al. Measuring the fitness of fitness trackers. In: Sensors Applications Symposium 2017 IEEE; March 2017. p. 1—6.

Clinical trial app regulations

13

Urs-Vito Albrecht, MD, PhD, MPH

Senior Scientist, Peter L. Reichertz Institute for Medical Informatics of TU Braunschweig and Hannover Medical School, Hannover, Germany

The need

A software application is defined as standalone software. If the application is intended for diagnosis or therapy, it will become a medical device in the legal sense. Examples could be software applications for the presentation of a patient's heart rate or other physiological parameters during routine checkups or for intensive care, monitoring in general, or software for measuring relevant parameters in clinical research.

In this case, the manufacturer has to make sure that he observes all applicable regulations for medical devices. For the European Market, the European Commission provides the manufacturer with information on whether an application is classified as a medical device or not. The MEDDEV Guideline 2.1.1/6 (dated July 2016) is legally not binding but very helpful for interpreting the appropriate European regulation. In contrast, the Medical Device Regulation (MDR) adopted on April 5, 2017, is a legally binding regulation for the member states. Originally, it was to have been enforced on May 25, 2020, but due to the SARS-CoV-2 pandemic, this was postponed until 2021. The MDR is much stricter in its requirements and redefines the risk classes that apply to health software. Only a very limited number of apps will still fall into a low-risk class in the future. The regulatory effort will definitely increase.

The solution

The consequences of the previously raised issues underline the necessity of obtaining a CE label for an app that is rated as a medical device before the manufacturer is allowed to put the application on the market—or uses it in a clinical trial. The CE mark may only be assigned after the appropriate conformity assessment procedures. The details of the procedures depend on the potential risk of the medical device.

For the US market, the FDA defines a medical device as,

an instrument, apparatus, implement, machine, contrivance, implant, in vitro reagent, or other similar or related article, including a component part, or

Innovation in Clinical Trial Methodologies. https://doi.org/10.1016/B978-0-12-824490-6.00002-5

accessory which is recognized in the official National Formulary, or the United States Pharmacopoeia, or any supplement to them, intended for use in the diagnosis of a disease or other conditions, or in the cure, mitigation, treatment, or prevention of disease, in man or other animals, or intended to affect the structure or any function of the body of man or other animals, and which does not achieve its primary intended purposes through chemical action within or on the body of man or other animals and which is not dependent upon being metabolized for the achievement of any of its primary intended purposes [1].

The manufacturers need a premarket notification if the device underlies the regulation. A letter of substantial equivalence from the FDA is necessary. Without obtaining this, manufacturers are not allowed to commercially distribute their devices. Originally issued in 2013, the FDA published a guideline for developers of mobile medical apps, which was last updated on September 27, 2019 [2]. Its intent is comparable to the MEDDEV guideline in the European Regulation, provided by the European Commission, and the document explains how the agency plans to exercise its oversight of device software functions (also covering mobile medical apps), with a focus on higher-risk applications. If apps are used in clinical studies, their use has to comply with the fundamental rules for clinical research, e.g., the Declaration of Helsinki or the ICH-GCP.

For participants providing their data for research, data protection is even more important, as they share sensitive and highly personal information related not only to their health, but also about their daily living routines, as well as information about their environment and interactions. From the ethical perspective, this results in the responsibility of the beneficiaries, namely the researchers, to establish the best possible measures for data protection to avoid unintended and unwanted data sharing with third parties. Policies for data security and privacy differ between countries but, with its General Data Protection Regulation (GDPR, adopted on April 14, 2016, gone into effect on May 25, 2018), the European Union laid the groundwork for harmonizing rules on data protection within the member states and their national laws on data protection [3]. It has replaced the Data Protection Directive 95/46/EC from 1995 with the intention to strengthen the digital privacy rights of EU citizens, improve the EU's online economy, and strengthen the free market. The regulation prohibits the processing of sensitive personal data including "data concerning health" that comprise "personal data related to the physical or mental health of a natural person, including the provision of health care services, which reveal information about his or her health status" (Article 4, No. 15 [3]). The regulation also allows exceptions for data processing for research and statistics. Article 9 on "processing of special categories of personal data" gives further insight into possible exemptions in this context [3].

Besides other changes, the key element of the GDPR is that individuals must give explicit consent for data to be processed (Article 6, No. 1 (a) [3]). Also, individuals will be granted easier access to their data, and with the "right to data portability," they will be enabled to more easily transfer their data from one service provider to another (Article 13, No. 2 (b) [3]). A sponsor has to make sure that the specifications of the GDPR are observed. He also needs to ensure that, as the GDPR is a regulation, not a directive—and individual member states are therefore allowed some flexibility regarding certain aspects—such adaptations are also respected.

Take-home message

Clinical trial apps must follow the highest possible standards in development in all relevant areas (data acquisition and management, data protection). It is also important to provide a transparent and detailed standardized description of the functionality of the app and to specify the rationale for the trial. Regulatory oversight is highly likely as clinical trial apps are a medical product.

The General Data Protection Regulation massively affects the clinical trial app approach: while it makes provisions for risk-based acquisition and processing of data, research may still be hampered as specific modalities for research can be adapted (Article 89 [3]) by individual member states of the EU [4]. There are also concerns that, due to its complexity, the GDPR is often overinterpreted by those without appropriate GDPR training [4].

References

[1] FDA, Federal food, drug, and cosmetics act, chapter II definitions, sec.201. [21U.S.C.321].
[2] FDA. Policy for device software functions and mobile medical applications. Guidance for industry and food and drug administration staff. 2019. Document issued on September 27, 2019, https://www.fda.gov/media/80958/download.
[3] Regulation (EU) 2016/679 of the European Parliament and of the Council of 27 April 2016 on the protection of natural persons with regard to the processing of personal data and on the free movement of such data, and repealing Directive 95/46/EC (General Data Protection Regulation) (Text with EEA relevance). https://eur-lex.europa.eu/legal-content/EN/TXT/?uri=CELEX:02016R0679-20160504.
[4] Negrouk A, Lacombe D. Does GDPR harm or benefit research participants? An EORTC point of view. Lancet Oncol 2018;19:1278−80.

The trial site

Brendan M. Buckley, MD, DPhil, FRCPI [1], **Frank M. Berger, MD** [2]

[1]*Chief Medical Officer, Teckro, Limerick, Ireland;* [2]*Head of Data Analytics, Global Clinical Operations, Boehringer Ingelheim, Ingelheim, Germany*

The need

Find a site

It remains a fundamental challenge in all trials to identify, contract, train, and support good trial sites. This will be the case even if trials in future use fewer sites and rely more on technology to collect data directly from study subjects. Sites must be capable of recruiting to target, of retaining study subjects, and of conducting the trial to a high standard of compliance.

A fundamental issue in attempting to ensure high standards of performance at trial sites lies in the fact that many trials, especially from Phase IIb onward, are conducted mainly by investigators whose primary job is diagnosis and appropriate treatment of patients, rather than to conduct clinical studies. With the exception of specialized commercial sites, trials are generally a voluntary addition to the core job of most investigators. For this majority of sites, conducting clinical trials is a peripheral task that is subsidiary to clinical service delivery, with its pressures and regular crises as currently exemplified in by the COVID-19 pandemic. These always take priority over trial-related activities.

The cost of initiating a site is estimated at about $10,000 to $30,000, whether or not it subsequently recruits any subjects. Although sites that recruit nobody may be shut down, those that recruit very few subjects generally need to be maintained, at an estimated cost of around $2000 per month. Traditionally, investigator sites are invited to participate in trials based on databases that reflect the accumulated experience of the sponsor or Clinical Research Organization (CRO). These are generated from sites previously employed and from local knowledge of hospitals or clinics known to be interested in clinical research. Recruitment websites are often used as entry points for new sites into these databases. Additionally, local franchises of the sponsor company may involve local key opinion leaders as high-profile investigators for marketing purposes in case the product gets on the market. A number of service companies exist that focus on identifying investigators by accumulating data from publicly accessible databases such as PubMed and clinicaltrials.gov and specialty lists of senior clinical staff on hospital and university websites.

Innovation in Clinical Trial Methodologies. https://doi.org/10.1016/B978-0-12-824490-6.00014-1

Patient organizations are an important source of intelligence through their websites and otherwise, especially for rare diseases and orphan drug development. There has been considerable interest in the use of social media as a means of investigator and site recruitment. Its actual use is slowly but surely increasing, with less than 3000 hits on searching the term "clinical investigator" in Linkedin in 2014, up to 6500 in 2020.

Train a site

Investigator meetings have been the usual way of trying to train investigators and site staff to perform specific trials for decades. These have never been shown to be effective and there are many reasons to believe that they are largely a waste of money (also see Fig. 14.1). Investigators have the problem of taking time off from clinical service to attend, particularly if significant travel is entailed. Many trials with which we have been involved have gathered less than 20% of primary investigators together and the meetings have been attended mainly by site coordinators as well as by subinvestigators whose trial involvement may be short-lived. Meetings are often held months before many sites will see their first screen. By the time the first subject appears, it is likely that little of what has been communicated at the meeting will be remembered. Meetings usually attempt to pack large quantities of information into a short time: the term "Death by PowerPoint" has been used many times about the format. Typical meeting components, for instance detailed instructions about handling blood samples for the central laboratory, are almost impossible to

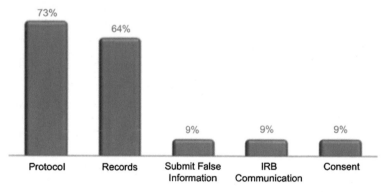

*Based on letter issued date. Inspections may have multiple deficiencies. Includes OAI untitled letters. (OSI database as of January 31, 2014).
Note that this does not denote number of inspections completed, but rather number of inspections report evaluated and closed in FY2013.

FIGURE 14.1

Frequency of clinical investigator—related deficiencies based on post-inspection correspondence issued. It is evident that sites lack fundamental understanding of the protocol and the specific administrative requirements of a clinical trial.

remember even immediately afterward. There is probably some merit in being able to network with trial sites at such meetings but even that may be frustrated if they are held in an attractive location that lures some of the participants away to see the sights.

Lack of effective site training is a major contributor to problems in trials, and there are many ways in which this may contribute significantly to failure of a drug's development. A basic reason relates to the power calculations, which are the fundamental mathematical expression of a trial's design. This calculation assigns a predicted or assumed variance to the measurements that are being compared between active and comparator groups. If the trial sites together do not make these measurements exactly as the protocol specifies, the variance is likely to be greater than that assumed in the design and the trial may fail to show a significant difference between the trial treatment and the comparator (Fig. 14.2).

The solution

The Internet provides a powerful means to train and support trial staff. This can be self-paced, targeted at different roles on site, and available throughout the trial to accommodate new staff starting midtrial. However, many training providers simply present slide shows and "talking head" videos on the Internet and these are not the answer. Web-based training that can be run on a computer in the background, minimized and silenced, while the "trainees" check their e-mails, is merely a futile token gesture. At a minimum, training should be intelligent, highly interactive, concentrated on problem areas, and focused on getting trainees to solve trial-relevant scenarios. To allow certification of users, it should be evaluated and tracked. It is

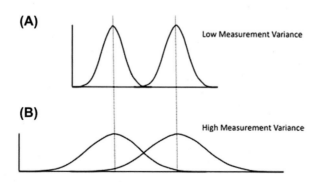

FIGURE 14.2

Effect of measurement variance on a comparison of two treatments. In (A), the variances in both treatments are low and there is a clear difference between them. In (B), the variances are large, for instance, because of differences in how well and precisely individual trial sites make the measurements, so the treatments do not appear to be significantly different even though the means are the same in both circumstances.

common for sponsors to allow web-based training systems as well as investigator meetings to be optional and to rely on staff training signoff at site initiation visits. This cannot be in keeping with the spirit of Good Clinical Practice (GCP) and the Declaration of Helsinki.

Surprisingly, given the importance of study site staff competence, little has been published on the formal evaluation of training methods, probably reflecting complacency with the status quo despite all evidence. A fairly simple correlation between the number of sites' interactions with an online training and support tool and the occurrence of protocol deviations points to a definite effect of the tool on quality. Fig. 14.3 shows in one trial how sites' interactions with a study-specific training and visit-by-visit guide influence their protocol deviation rate, demonstrating that about 80% of deviations are committed by the sites having the lowest quintile of usage.

Trial protocols are typically subject to amendments, which in themselves present a significant challenge to training and compliance. The continuing standard use of paper protocols, that need to be replaced with every amendment, must contribute to the dominance of protocol deviations in adverse inspection findings (Fig. 14.1). In the COVID era, having trial staff in personal protective equipment and the protocol in an office, either as a paper document or as a pdf on a desktop computer portal, means that it cannot realistically be read when it is needed. COVID has emphasized that protocol and other trial information on mobile personal (not shared) devices is an essential safe support for trial sites.

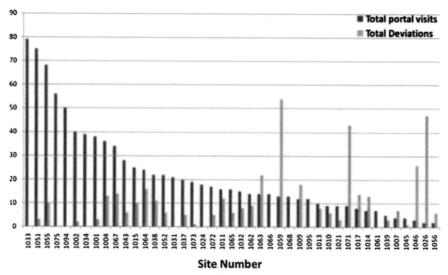

FIGURE 14.3

Effect of frequency of interactions by sites with Firecrest study-specific training and support portal (in red) on the rate of protocol deviations in each site (green).

Data courtesy of Firecrest Clinical Ltd (an ICON pl company).

Even with the best training, the reality is that many trial sites see only a few trial subjects per month and often do so while conducting other studies in parallel. It could be said that existing approaches to investigator trial site selection are based on the application of business processes rather than of science. They largely entail the aggregation of information from disparate systems that are accessed more opportunistically than systematically. The process of engineering a better way of selecting sites should entail "(t)he creative application of scientific principles to design or develop structures, ... processes, ... utilizing them singly or in combination; or to construct or operate the same with full cognizance of their design; or to forecast their behavior under specific operating conditions" [1].

Such an analytical approach to site selection would seek to utilize the enormous quantity of data that exists about study sites and patients in various repositories. Among the most important of these are electronic medical records. As more institutions move to electronic medical records, there should be a great opportunity to mine these to discover valuable data about demographics of disease distribution, where diseases are treated, who treats them, how they are treated, and what are the outcomes. Access to medical records is regarded as highly sensitive and is tightly controlled. In the United States the "HIPAA Act" [2] and in the European Union the General Data Protection Regulation [3] strictly limit access to records. Nevertheless, there are solutions in place to use this resource in a manner that is fully compliant with individual privacy rights and the law, e.g., through companies like TriNetX or Clinerion.

Already, Pharma companies, CROs, and electronic data capture (EDC) providers control very large trial databases. Information about trial sites' activity is often buried in these, such as their therapeutic area competences, human resource numbers, and recruitment metrics. In addition, there is valuable information present about performance quality, for example, the rates of queries raised by monitors, Serious Adverse Event (SAE) reporting compliance, laboratory sample requirement compliance, and protocol deviation and violation rates are all present and to be found in study files. Using these sources and employing a "big data" approach, it should be possible to test a series of hypotheses on the correlation of activity with performance as well as with demographics.

The role of analytics

Nowadays more and more pharmaceutical companies ask for robust benchmarking data, both external and internal (if available), as reference figures for planning processes. This could be the number of eligible patients under a certain set of inclusion and exclusion criteria, realistic recruitment rates, screening failures rates, etc. The desire is to move away from "common beliefs" to sound decisions based on sound data. A sponsor's own historic data may be insufficient to provide these insights, either simply due to the size of the company, or due to lack of suitable data, e.g., caused by a shift in focus regarding the targeted therapeutic areas. As a result, we can observe a substantial growth in numbers of companies offering either access to such industry-wide metadata.

Pharmaceutical companies (including CROs) are making progress along the continuum of Business Analytics capabilities:

A. Descriptive Analytics—What happened?
B. Diagnostic Analytics—Why did it happen?
C. Predictive Analytics—What will happen?
D. Prescriptive Analytics—What should we do to make it happen?

This journey requires suitable data to work with. It is highly desirable to use a holistic approach ("cost and conduct"), integrating operational metadata (e.g., reflecting recruitment progress) and financial data (e.g., actual cost spent until a certain milestone). Equally essential are qualified personnel—either in-house or contracted—with the necessary analytical skills. Many companies now establish in-house Data Science teams to improve the analytical capabilities in all areas of the business, not only in Clinical Operations.

In our specific challenge of identifying the 'right' sites, modern analytic methods, in particular Machine Learning, can be used to develop a predictive model of whether a site candidate will be recruiting study subjects in time and in compliance with the protocol [4]. The amount of available data to train such a model should already be sufficient and is certainly increasing steadily.

Two main challenges are to be overcome:

• Resistance against this new approach and its results and insights "I have always done it this way, why should I now believe the computer to do my business differently?"
• Identifying and recruiting personnel able to bridge between both worlds: the world of developing new medications, and the world of data handling, statistics, predictive modeling, and data visualization.

Beyond the subject of site selection, here are some other key objectives of such analytics:

For trial preparation, it helps to understand the interdependency of sample size (# of subjects), # of sites, timelines, and recruitment rate (# subj/site/month). Currently, attempts are made in the context of ICH E6 R2 and Quality-by-Design (ICH E8 R1) to better quantify trial complexity and protocol complexity. With access to such parameters, analytics are playing a growing role in feasibility and site selection.

When monitoring trial progress, the quantification of any risk is of utmost importance and a requirement based on ICH E6 R2/GCP. This includes the understanding of the causal chain to successful recruitment—and using this understanding to implement a harmonized approach how to analyze if you are behind in recruitment.

Cost planning (budgeting) and tracking also benefit from more evidence by differentiating assumptions and translation of those assumptions into cost and cost drivers.

Swot analysis

Strengths: Sites are primarily expected to deliver good-quality work. Big data plus more interactive training should enable selection and preparation of the best possible sites.

Weaknesses: Not all useful data on sites are always available because of data protection limitations. Not all scientifically competent sites are interested in applying the appropriate technologies to increase their trial competence.

Opportunities: Economic pressure on sites and sponsors will enforce more professionalism in trial conduct. More and more data is becoming publicly available, and the amount of IT investment required to run even sophisticated predictive models is decreasing.

Threats: All data available are historic, and thus may no longer reflect the current site status. Access to data in near real-time is needed.

Take-home message

Finding, training, and supporting trial sites has got to leave behind systems that are demonstrably ineffective and wasteful in favor of a systematic and comprehensive process based on the application of modern instructional design and communications technology. It is time to rethink fundamentally the process for instructing site staff, for maintaining and growing their competence throughout the trial, and for supporting them with modern mobile access to trial information. We owe it to trial subjects for their better protection and to the proper use of investment that we enable new treatments to be given an error-free and fair trial.

When clinical operations wants to move to a more data-driven organization, the problem is not about having enough data—the data are there—it is about the extraction of meaningful information from the exponentially growing amount of raw data, and using these data-derived insights for business decisions to boost the efficiency of conducting clinical trials. Data analytics helps to identify patterns in historic data that help predict future outcomes (recruitment, timelines, cost). Data analytics play an essential role in standardizing metrics and KPIs and the way we look at these figures in a harmonized, user-friendly way, thus enabling data-driven business decisions. We need to ensure that the various education pathways leading toward a career in the pharmaceutical industry do not only include scientific training and project management skills, but also a generous helping of data analytical and visualization skills.

References

[1] Engineers' Council for Professional Development. Canons of ethics for engineers. New York: Engineers' Council for Professional Development; 1947.

[2] Health Insurance Portability and Accountability Act. Public law 104-191. 104th Congress of the United States of America. 1996.

[3] European Commission. Regulation (EU) 2016/679 of the European Parliament and of the Council of 27 April 2016 on the protection of natural persons with regard to the processing of personal data and on the free movement of such data, and repealing Directive 95/46/EC (General Data Protection Regulation).

[4] IQVIA White Paper. AI in clinical development. 2019. https://www.iqvia.com/-/media/iqvia/pdfs/library/white-papers/ai-in-clinical-development.pdf. [Accessed 27 May 2020].

A single-center "virtual" trial with patients across the USA

15

Miguel Orri

AlgaeVir Sdn Bhd, Petaling Jaya, Selangor, Malaysia

Foreword

The term "virtual trial" is also used for exercises conducted entirely in silico [see section by L A Ogilvie et al. "In Silico Trials"]. When used in connection with any type of study in which patients are dosed with an IP, it may be misleading, because it still is a real and not "virtual" trial. Others favor the term "Decentralised," but that leads to the question "What is 'the centre' from which the trial is removed ?"

These terminologies may consolidate further as people become more rigorous in describing novel data-gathering methodologies. However, since very commonly in use, in this section the described trial is characterized as "virtual."

The need

Participant recruitment is often a major barrier to the feasibility and timely completion of clinical trials. Trial participation is often limited by geographical constraints due to the number of eligible participants who live near trial sites, especially for uncommon medical conditions. In addition, a requirement for multiple visits to the clinical site and a limited number of time slots can create a funnel for trial enrollment and interfere with completion. As we have learned during the COVID-19 pandemic, travel is a risk in itself and may not always be possible at all.

This dilemma may be enhanced by the fact that patients are not always fully engaged in clinical trials, and therefore patient data are not optimally obtained or processed. Although considerable attention is still focused on monitoring of medical records and the transcription of those into case report forms, the information obtained from the primary source, the patient, could be improved through better patient engagement, particularly in the light of increasing complexity, which makes clinical studies burdensome for patients. Also see section by B Buckles: "The Patient as Sub-Investigator."

Furthermore, the primary care physician (PCP), often the patient's first and most trusted point of contact for healthcare, is mostly not incentivized to refer patients to clinical trial centers and in most cases insufficiently informed about the treatment response of their patients.

Innovation in Clinical Trial Methodologies. https://doi.org/10.1016/B978-0-12-824490-6.00004-9

These challenges as well as increasing demands on site personnel lead to a decreased interest from investigators in trial participation and maintaining consistent quality across multiple sites can pose additional challenges for study teams. The operation of trial sites (i.e., site management, payments, and monitoring) accounts for most of the costs of conducting clinical trials [1]. The current clinical trial paradigm does not allow sustainable innovation, and a more efficient way of conducting clinical trials is needed to reduce their current cost, improve timelines, and optimize data quality.

In parallel with these challenges, the proliferation of health information on Internet is giving rise to the "e-patient," a patient who is engaged, empowered, and online (also see P. Kolominsky "Patient-centric Registries"). The widespread use of mobile phones has decreased socioeconomic and age barriers to web access [2]. Groups of patients with common medical conditions and concerns are also organizing online to share data and initiate research activities [3].

The solution

Participatory patient-centered (PPC), web-based clinical trials could increase patient access to clinical trials, streamline trial conduct, and facilitate rapid reporting of clinical signals and outcomes to regulators, healthcare providers, and patients.

The key components of PPC web-based trials are listed below.

Subject

- Web-based recruitment
- Web-based multimedia informed consent process
- Web-based screening

Technology

- Mobile communication device-based efficacy assessments (e.g., e-diary)
- Interactive remote data capture via secure patient portal
- Real-time data access for sites, monitors, and auditors

Investigator site

- Coordinating function for virtual assessments: patient does not attend investigator site
- Study drug delivery by overnight courier
- Study physician/call center available 24/7 by e-mail and/or phone
- Real-time safety data processing
- Study data returned to subject

More rapid recruitment and the potential to reduce clinical site costs are strong incentives for pharmaceutical companies to conduct web-based trials. Easier access and convenience will enable more patients to join PPC web-based trials.

The **R**esearch on **E**lectronic **M**onitoring of **O**veractive bladder **T**reatment **E**xperience (REMOTE) study was the first study conducted within this innovative framework with an investigational new drug (IND) application under the supervision of the Food and Drug Administration (FDA) [4]. Although this study assessed a variety of novel concepts and new technologies for data capture and handling, one of the main aims of this study was to demonstrate that it can be conducted in a completely remote manner. Its modular concept allows for a stepwise improvement of the clinical trial processes for a large number of clinical trials in all phases of clinical development.

In the REMOTE study, which was based in the United States only, potential participants were recruited using a variety of web-based sources, including targeted advertisements on online search engines and health websites; online communities with a potential interest in target disease; online patient advocacy groups, social media sites, and Craigslist, as well as from healthcare organizations, community-based clinics, and commercial recruitment vendors.

Potential participants were initially screened for eligibility using web-based questionnaires and laboratory test results. Additional precautionary procedures were included to verify participant identification and screening. To minimize potential fraud, approval was obtained from eligible participants to perform a secure and confidential third-party online identity verification using personal information. The online identity verification process did not include financial or credit checks but used a public records-based service widely used for online identity confirmation. To prevent hacking of trial websites, an e-mail confirmation process was required wherein a time-sensitive link was sent to the potential participant's e-mail address. Once e-mail confirmation and identity confirmation were completed, the participants completed the online informed consent process followed by screening questionnaires based on study eligibility criteria. Those who signed the informed consent document and met screening criteria were contacted by the investigator's study staff for a telephone discussion about the trial and review of the informed consent details. After these telephone calls, appropriate patients were countersigned into the study and scheduled for laboratory testing and physical examination. After the study investigator reviewed the test results, those who remained eligible were enrolled in the placebo run-in phase. At this point, participants were sent single-blind study medication by courier to their home. Participants also received a mobile feature phone with a custom application installed for entering e-diary data. Participants who remained interested and eligible were randomized via an interactive voice response/interactive web response system to double-blind treatment and followed-up with mobile phone-based diaries and web-based questionnaires. Patients had access to a physician from the investigator team at all times.

Physician engagement

In the REMOTE study, a national clinician network was used to do the physical examination and to provide a report to the investigator for his or her evaluation, similarly as this is often done with electrocardiograms in conventional studies. For many late-phase studies, such tasks including blood or urine sampling and simple safety follow-up examinations, which do not need study-specific training, could be performed by the patient's own PCP without making them investigators. This would keep the PCPs closely connected with their patients' treatment and up to date with the clinical trial progress while getting reimbursed for their services.

From a study site perspective, this new paradigm makes the conduct of clinical trials much more efficient. As protocols become more and more complex and requirements for training and documentation compliance become more burdensome, it can be difficult for site personnel to maintain their focus on a small number of participants on a part-time basis while adhering to all protocol requirements. This new paradigm allows sites to become coordinating centers and dedicate staff to recruit large numbers of patients in a short timeframe, and a lot of the documentation can be entered by the participants themselves directly into the system. As clinical trial platforms become commercially available, study sites can use a single platform for the management of several studies as these get added in a modular manner, and at the same time most of the training and documentation could be handled by one system, which would enable sites to work more efficiently.

Recruitment

Potential subjects may be identified through many different sources that may be available on a local, regional, national, or study-wide level. Although most studies still rely on small investigator databases and several different advertising modalities, the virtual trial model offers itself up to web-based advertising, collaboration with large patient organizations, and regional or national health databases as they become more and more available.

To engage potential participants from the beginning, it is essential to make the recruitment process as patient friendly as possible. Long web-based questionnaires over multiple screens may not be the best way even if a helpdesk is at hand. At the early stage of recruitment, it is paramount to gain the interest of potential participants in a trial, which may be suitable for them, by providing the essential facts about the study and demonstrating that their concerns are being addressed, allowing them to identify with the study. It is self-evident that all the information provided does need ethics committee or institutional review board (IRB) approval before it is presented to potential trial participants.

Another way of making the process more patient friendly and efficient, by reducing duplication of work in determining the eligibility of a potential participant, may be to engage a recruitment vendor who can elicit patient eligibility using an approved study questionnaire over the phone and entering the data directly into the study database that can then be verified by the investigator. In this way, the burden is taken away from the participant.

Identity verification

Although in conventional trials, the patient is often known to the investigator or at least the existence of a real person as well as the age group and sex can easily be identified, this poses a new challenge for virtual trials where additional steps are needed to ensure the integrity of the trial.

At least in the United States, there are vendors who use information from publicly available databases to verify a person's identity based on questions that can only be answered by the individual. This system is also used by banks and government agencies and allows for a high level of security. However, it requires a participant to enter personal information such as a national insurance number and address to be identified, and many participants might not be willing to enter such information over Internet specially at a stage when they do not even know whether they are eligible for the study. Furthermore, there might not be enough data publicly available for a person to be identified, or the participant might get answers wrong, which could automatically exclude the person from participation.

As it is important to ensure that the person who signs the informed consent is actually the person participating in the clinical trial, it is essential that this link can be established and documented. This may be accomplished through the use of a trusted healthcare provider such as the patient's PCP, a pharmacist, or a local laboratory who may be engaged in the clinical trial.

Informed consent process

To accommodate the virtual nature of a clinical trial, the informed consent process needs to be well thought through and may necessitate several steps depending on the type of research and potential risks for the participants. However, to enhance recruitment and retention and to provide better-quality data for clinical trials, it is important to have a patient-centric informed consent process that leverages multimedia to thoroughly and consistently explain a study and to achieve a more knowledgeable and engaged patient. At the same time, this process needs to be well documented and compliant with International Conference on Harmonization good clinical practices and acceptable to regulatory agencies, IRBs, and ethics committees. Ideally, this is handled by a centralized process, enhancing the ease of retrieval of source data for inspections, audits, and monitoring.

In the case of the REMOTE study, each participant began the informed consent process by viewing and listening to an online automated slide presentation, in which the principal investigator described the essential elements of the trial, followed by a review of the full informed consent document online. Participants could print a hard copy of the informed consent document for easier viewing. Before signing the informed consent document, each participant was required to pass a multiple-choice test confirming the individual's understanding of the informed consent document. After the participant passed the informed consent document test and read and signed the informed consent document, each participant also received a phone call from study personnel who discussed the informed consent document with the participant, answered any questions about the informed consent document or the trial, reviewed subject-reported medications and medical history as appropriate, and addressed any concerns. If the individual fulfilled study eligibility requirements and remained interested in participating in the trial, the study investigator counter-signed the informed consent document. Two hard copies of the electronically or hand-signed informed consent document were then mailed to the participant. One copy was for the participant's records and the second copy was to be given to the PCP or health professional, should they seek medical care during the study.

This process provides a real-time tracking option, a more consistent and standardized process across sites where both the information provided to subjects and the resultant understanding are well documented and allow for real-time remote monitoring and audit access. The flow of the informed consent process is illustrated in Fig. 15.1. This process has also been acknowledged by the US Department of Health and Human Services in a document on Considerations and Recommendations Concerning Internet Research and Human Subjects Research Regulations [5].

As the computer technology is developing at a fast pace and an increasing number of informed consent solution providers enter the market, more patient friendly tablet or even downloadable application-based solutions are nowadays available not only for virtual trials but also as a better informed consent process for conventional studies. Outside the United States, the consent cannot always be provided electronically and needs to be assessed on a country by country basis, but the study information can be provided electronically in most countries globally.

Dispensing study medication

The dispensation of study medication also needs careful planning in the virtual setting, and country-specific laws and regulations require consideration especially for investigational new drugs (INDs). In the United States, for example, federal regulations state that study medications under an IND application must be shipped to study investigators for administration only to participants who are under their direct supervision, an investigational product waiver of this requirement of 21 CFR 312.61 under 21 CFR 312.10 [6] may need to be requested by the sponsor of a clinical trial. Other countries may have requirements for drugs to be shipped from a depot or through a pharmacy. The chain of custody needs to be clearly documented, and some additional steps from the participants side are needed to ensure the safe delivery and use of study medication. For example, once the study investigator has authorized the dispensation of trial medication possibly through an interactive response system, it may be sent to the participants' physical address using an overnight delivery service with a signature requirement. The participant then needs to confirm receipt of trial medication and the condition of the contents via the study computer portal. Receipt of this confirmation may trigger the start of a treatment phase. In each phase of the study, used and unused study medication bottles can be returned by each participant to the investigator site by courier using a preaddressed stamped envelope. Once received at the investigator site, study drug will be subject to the usual reconciliation and destruction process.

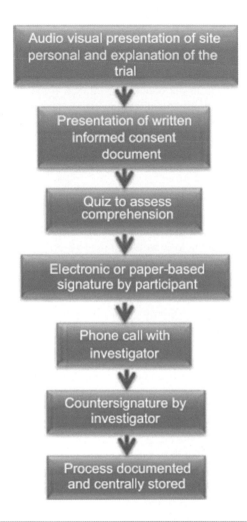

FIGURE 15.1

The electronic informed consent process in the REMOTE study.

Web-based computer system

One of the key elements of the virtual trial setting is a computer portal designed for electronic data capture directly from the participant who may enter consent and study related data as well as for the data captured at the assessments throughout the study. This may include patient reported outcomes, and data from other linked-up devices such as glucometers, blood pressure cuffs, or weighing scales, to name just a few. These data will be accessible to the investigator in the portal, and the investigator will review these to assess the eligibility or the progress of a subject. The review of screening data may also include medical history, physical examination, and laboratory reports from the PCP, which may be entered into the portal.

The investigator can review all these data, and this review may also be documented in the portal. As these electronic clinical trial platforms are developing at a very fast pace, newer systems allow for patient recruitment, site management, clinical trial management, and reporting, all to be performed on the same platform allowing for even more efficiencies in the conduct of clinical trials. It is important that the computer system is compliant with all applicable guidances and regulations, appropriately tested and validated, and retained in accordance with relevant record retention requirements.

Electronic data capture

The electronic remote data capture is another key element of virtual trials that may be used in a large number of conventional clinical trials to improve data quality and accessibility. It requires some upfront investment of time and resource to define the system requirements, similar to a well-designed electronic case report form. The fundamental difference is that participants can enter data directly into the system with the option for logic checks at the time of data entry; this capability may substantially improve data quality and reduce the need for monitoring of source data at the trial site, leading to data queries, as the data have been entered directly into the system by the participant, reducing the risk of data transfer errors. See also section "The Patient as Sub-PI" from B Buckley.

Any remaining data queries can be addressed in real time as the data can be monitored as soon as they have been entered, enabling the investigator and sponsor to get faster access to safety signals on a patient and study level. The real-time access to the data by the investigator also allows the site personnel to follow-up on missing data immediately and therefore potentially improve the data quality. It may also lead to substantial time saving at the end of the study as database lock may be achieved soon after last subject last visit rather than several weeks later as is often the case in conventional studies.

The data flow for a virtual trial model is illustrated in Fig. 15.2.

Safety monitoring

As in other trials conducted by community-based research sites rather than physician practices, the Clinical Trial Coordinating Center in a virtual study provides medical care for the subjects in the trial by referring subjects to their own PCP, a specialist, or emergency care as appropriate. This should be explained to the participants in the consent document.

As in conventional clinical trials, all adverse events are reportable from the time that the subject provides informed consent, either directly by the subject, the PCP, or any other treating healthcare professional.

The time of awareness of an adverse event by the sponsor is the time of receipt of the initial report at the call center or the computer portal.

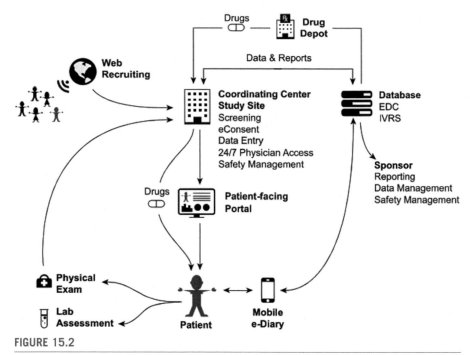

FIGURE 15.2

Electronic data collection (EDC) and management. *IVRS*, interactive voice response system.

All adverse events have to be followed-up as appropriate by the investigator to obtain information adequate to determine the severity, causality, and outcome of the event and to assess whether it meets the criteria for classification as a serious adverse event. It is therefore necessary to have a process in place that ensures the investigator has access and documents his review of all reported adverse events within 24 h.

Patient engagement

Although this concept enables the patient to participate without geographic limitations mainly from the comfort of their home and at the time of day that is convenient to them, it is not in itself a guarantee for patient engagement and compliance; therefore, it is essential that this subject is addressed separately. In the absence of a face-to-face interaction with the investigator, the computer interface used has to be simple, intuitive, trustworthy, and suitable for the study population. An Internet-based study website that requires the use of a web browser with several access steps may not be suitable for an elderly population, particularly in less-developed countries, whereas a study in adolescents might offer itself up to a game-like application to engage participants in the flow of the study. To make the system as easy as possible for the participant to access, computer tablets may be the best solution as they can be preprogrammed to allow for access with the push of a single button while the connectivity to Internet or Cloud may be established in the background. Although currently it may be advisable to provide participants with a preprogrammed device, one can well imagine that with increasing penetration of smartphones, a downloadable clinical trial application might be a viable alternative for virtual clinical trials.

Apart from the technical aspects, an effort should be made to keep the participants informed of their progress to encourage them to perform all study-required tasks as needed. Again, the direct real-time access to the subject via a tablet or phone-based communication device allows the investigator team to remind the participants of upcoming assessments or request any missing data in a timely manner while reinforcing the message that their contribution is important and appreciated. The device can also be used to provide participants with information about the study or their disease, but provisions should be made to give the participant control of how much information they want to receive.

Each study population will respond to different ways of engagement, and this needs to be considered at an early stage of the trial design.

Finally, it is part of this model to provide individual study data back to participants at the end of the study in an electronic format, which would allow them to share the data with their healthcare provider and to add it to their personal health records.

Generalizability

Completely virtual trials, at this point, may only be suitable for a limited number of late-phase clinical trials where the safety profile of a medicine has been established. At the same time, it is those large postapproval Phase IIIb and IV trials where the biggest efficiencies can be achieved. There is also an increasing demand for real-world data, necessitating more of these large trials, often requiring 10,000 patients or more, which can be facilitated by the virtual trial setting with the opportunity of realizing substantial efficiencies. The regulatory authorities, such as FDA and some European agencies, have been supportive of this innovative approach and generally encourage the pursuit of innovative and more efficient ways of conducting clinical trials.

Although some of the above-mentioned considerations focused on clinical trials under an IND application, most principles equally apply for trials with approved drugs and many elements of this model might be even more attractive for noninterventional trials.

It needs to be emphasized that the virtual trial model by no means is an all-or-nothing decision. Many of the individual elements can be incorporated in a large number of clinical trials at all stages of development to improve data quality, patient engagement and efficiency. Prime examples are the improved informed consent process and the remote data capture directly from the participant with the benefit of real-time data availability.

SWOT analysis

Strengths: Virtual trials are designed for those most relevant—the patients. This has a positive impact on recruitment, retention, and data quality. The concept enables trials with a reduced number of sites and remote monitoring which can lead to substantial cost savings.

Weaknesses: Not every trial is suitable to be conducted completely virtually. Overall complexity in planning at the sponsor end may initially be higher while the teams become familiar with the new electronic clinical trial platforms.

Opportunities: The growing population of "digital native" patients that are easily using modern communication tools and also seeking more independence in their lifestyle will prefer a "virtual" over a site-based study.

Threats: The lack of direct face-to-face contact with the investigator may not be the preferred option for elderly patients.

Take-home message

Virtual clinical trials are there to stay. Either in their entirety or as elements of a more conventional model, there is a potential for substantial benefits for patients, investigators, healthcare providers, and sponsors. Patient engagement is paramount for recruitment, retention, and data quality and needs to be thoroughly considered, both in studies with and without face-to-face interaction. The infrastructure and technology needed for virtual clinical studies is readily available, and the first virtual study has been successfully conducted. Regulatory authorities are supportive of this innovative approach as long as the fundamental patient rights, confidentiality, and safety are not compromised; however, the suitability of virtual trials needs to be assessed on a case-by-case basis. Even in a more conventional clinical trial setting, the available technology allows for substantial improvements in patient recruitment, consent, engagement, and retention; and data collection and handling as well as site selection and management can be substantially improved, potentially resulting in improved data quality, reduced timelines, and significant cost savings.

References

[1] Eisenstein EL, Lemons 2nd PW, Tardiff BE, Schulman KA, Jolly MK, Califf RM. Reducing the costs of phase III cardiovascular clinical trials. Am Heart 2005;149(3): 482–8.

[2] Smith A. Mobile access 2010. Pew Research Center's Internet & American Life Project; July 7, 2010. Available at: http://pewinternet.org/Reports/2010/Mobile-Access–2010. aspx.

[3] Wicks P, Vaughan TE, Massagli MP, Heywood J. Accelerated clinical discovery using self-reported patient data collected online and a patient-matching algorithm. Nat Biotechnol 2011;29(5):411−4.

[4] Orri M, Craig HL, Bradly PJ, Anthony JC, Steven RC. Web-based trial to evaluate the efficacy and safety of tolterodine ER 4 mg in participants with overactive bladder: REMOTE trial. Contemp Clin Trials July 2014;38(2):190−7.

[5] Considerations and recommendations concerning internet research and human subjects research regulations, with revisions final document, approved at SACHRP meeting. March 12−13. 2013. Available at: http://www.hhs.gov/ohrp/sachrp/mtgings/2013%20 March%20Mtg/internet_research.pdf.

[6] Investigational New Drug Application. Code of federal regulations title 21,vol. 5, Chapter I, Part 312 (21CFR312). Food and Drug Administration. Available at: http://www.accessdata.fda.gov/scripts/cdrh/cfdocs/cfcfr/CFRSearch.c:fi:n?CFRPart=312& showFR=1.

From data to decisions

Data standards as a pathway to interoperability

16

Wayne Kubick

Chief Technology Officer HL7 International, Ann Arbor, MI, United States

The need

Imagine how difficult it would be to sequence genomics data if science had not agreed to represent DNA using the fundamental CATG alphabet. In the pandemic situation, accelerated development and review of data for a new COVID-19 vaccine would not be possible without more streamlined data standards. Also in the case of nonvaccine clinical research data, which historically was captured on CRFs or in batch transfers of lab data rather than by tapping into data sources used in healthcare, an initial standard syntax and a set of controlled terminologies were provided though the Clinical Data Interchange Standards Consortium (CDISC) [1]. But that was only the start of a development for further standardization.

The solution

In July 2004, then FDA Commissioner Lester Crawford announced that FDA would accept CDISC SDTM datasets from sponsors in regulatory submissions [2]. Though the announcement did not specify that this would be a regulatory requirement, a series of notices of proposed rule changes that would allow FDA to require SDTM data was subsequently posted in the Federal Register, followed by a set of FDA Guidance documents on the topic [3].

In 2006 a Gartner report predicted that while total adoption of data standards from the start would eventually save billions, the return on investment would decrease by 60% if only adopted at the submission stage [4]. But the SDTM standard [5] was designed for representing tabulation data for the commonly used domains that were typically included with most clinical studies that had been traditionally sent to FDA as long listings in pdf format— and not necessarily optimized for capturing data on CRFs. To fully capitalize on the benefits of applying standards from the start sponsors would need structured protocol level information [6] and standardized CRF content—the latter of which became possible after the release of the CDISC Clinical Data Acquisition Standards Harmonization (CDASH) effort in 2008 [7]. Though, applying a set of standards at the start did not eliminate the need to convert over and over again on the path from protocol to submission.

Innovation in Clinical Trial Methodologies. https://doi.org/10.1016/B978-0-12-824490-6.00003-7

Data overload and quality

Data standards do not necessarily reduce the volume of data that can lead to data overload; however, the CDASH team recognized that they can certainly motivate researchers to avoid collecting unnecessary data [8], and substantially reduce the stress for the sites dealing with data when data conforms to known structures and familiar conventions. Certainly, most would agree that it is easier to read and rapidly comprehend something that is written in your native language, and CDISC data standards eventually became a common language of clinical research (Fig. 16.1). Data that conforms to known standards simply make researchers more productive, whether by allowing reuse of common building blocks in setting up studies or in enabling use of off-the-shelf review tools without the need for excessive configuration or in creating analysis programs based on standard code modules or finally in interpreting final results [9].

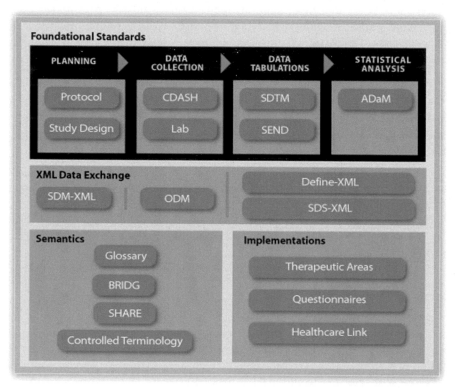

FIGURE 16.1

CDISC data standards.

Source: http://www.cdisc.org/standards-and-implementations — used by permission (a more recent version is available at https://www.cdisc.org/standards).

The impact of data standards on data quality has barely been scratched to date in the literature. One of the benefits of the full set of CDISC submission standards was to help FDA reviewers become more productive and minimize misinterpretations of the data [10]. Similarly within sponsors designing study databases with a clear, consistent end in mind should help improve operational efficiency by increasing reuse of software components, and help avoid common errors both by leveraging the learning curve and making it possible to use standard edit checks to identify errors sooner.

But, as the Gartner study realized, an even more fundamental benefit of standards on quality can be achieved when standards are employed at the earliest stages of study planning—in protocol and CRF design [11]. Also see Section "Data mining for better protocols" from Fareed Melhem to learn further reasons why data standards should be made available from the beginning of the trial process, i.e., the protocol.

When protocols are defined using the same standardized descriptions of collections of data elements, it should be possible to use these descriptions to set up data collection forms in a clinical data management system, which in turn could generate data listings that conform in general to SDTM datasets that provide inputs for analysis. But when the path to standards is through mapping from other formats, there is a risk of introducing new error and losing data fidelity and consistency across different sponsors, even if mapping transformations are, initially, the only realistic option toward standardization.

The FDA is well aware of this. While many of the current FDA guidelines focus on the end products they will receive [12], FDA has long recognized the critical importance of end-to-end traceability of study data, and FDA experts have contributed to efforts to develop structured protocol models. The need for standardized CRF representations was identified as number 41 of the important initiatives outlined in the FDA's 2007 Critical Path Initiative [13], which recognized that use of standard CDASH CRF representations can eliminate errors at the site, since site personnel are more likely to use familiar CRFs correctly and consistently with fewer misinterpretations.

It took some time, but the stick finally began to appear in 2013, which saw the issuance of a clear statement from FDA requiring use of CDISC [14]. Also in 2013, the European Medicines Authority (EMA), while not requiring data submissions, announced a commitment to support data transparency initiatives by requiring that research data be made available publicly, with the apparent intent of eventually requiring conformance to CDISC as well [15]. This was further enforced by statements from the Pharmaceuticals and Medical Devices Agency (PMDA), Japan, announcing a pilot and a plan to require use of CDISC standards by 2016 [16]. Recognition of the importance of using CDISC standards in all three regions of the International Conference for Harmonization (ICH) offered a compelling case for sponsors to accelerate their internal plans for implementing CDISC standards.

The reason for considering the use of clinical data standards as part of a data transparency initiative like EMA's is directly related to one of the key value propositions— data standards make it possible to pool, compare, and analyze disparate data sources in order to support knowledge discovery. This expectation, in fact, was recognized by Dr. Norman Stockbridge of FDA when proposing the creation of the Janus Data Warehouse in 2002 [17]—a project that had a significant influence on the development of the CDISC SDTM standard. By providing a framework where clinical data could be easily pooled together to support aggregate and comparative analysis, FDA reviewers could conceivably [18] improve their ability to compare the safety and effectiveness of new compounds to current standards of care, and more likely be able to continue to build aggregate safety databases that depict how a drug's safety profile evolves over time [19].

Of course, the ability to pool and aggregate data from different drugs or studies in a scientifically sound manner depends on more than structural data standards [20]. Since each clinical study protocol is unique, with different target endpoints, objectives, exposures, populations and constraints, it is essential to be able to provide context that can guide what data sources can be combined with integrity, which is the purpose of a structured protocol standard. In other words, first you must find which protocols are capable of comparison, and then you combine the data from those protocols before seeking knowledge from the pooled database. Second, while the structures help, it is essential also to have consistent controlled terminology so that the data values within the database mean the same. Because of the specific requirements of CDISC standards like CDASH and SDTM, which did not fit with the common terminologies widely in use among healthcare providers and payers, CDISC established a specialized controlled terminology initiative in partnership with the National Cancer Institute's Enterprise Vocabulary Services (EVS) [21] toward that end. Thirdly, it is most essential that different individuals develop a common understanding of how to use the standards, and to minimize individual judgments that can lead to inconsistencies in how data is represented— which will likely compromise the integrity of the pooled database. This is an evolving process involving better standards that close current gaps, better training and accumulated experience, improved validation tools and techniques to identify inconsistencies, and wider adoption among users of data standards.

But clearly, the ability to use standards consistently on the broad reservoir of current and past research data and make it available to a broader research community has the potential to drive dramatic new insights in the use of medical products to improve patient lives—even in some cases to potentially bring back past failed drugs and reapply them for new, safer uses.

Collaborative initiatives to develop standards

While the major regulatory authorities were still formulating their messages regarding data standards in 2012, other discussions relevant to clinical data standards were also occurring among a group of senior leaders for some of the world's largest pharmaceutical organizations, who were seeking to identify common areas of need where they could work together directly in a precompetitive manner. The benefits of such collaboration initiatives had previously been demonstrated by the Innovative Medicines Initiative in Europe [22], among other public-private partnerships. These discussions resulted in the establishment of TransCelerate Biopharma, Inc. (TCB) in 2012 [23]. TCB's initial portfolio included five major programs [24], one of which was to contribute to the development of therapeutic area data standards. The importance of applying data standards to specific therapeutic areas of clinical research has long been recognized—since it is those efficacy and disease-related questions relevant to a particular disease area, patient population, and strategic focus area where the greatest and most impactful benefits can be achieved. But it was the explicit mandate embedded in law through the FDAAA act of 2012 [25] that encoded this in law that finally caught the attention level of C-level scientific and business executives at major pharmaceutical companies. By mid-2012, TCB announced its participation to work with CDISC, FDA, and the Critical Path Institute [26] (a nonprofit public-private partnership) to participate in a therapeutic data standards initiative under the Coalition for Advancing Standards and Therapies (CFAST) [27].

While CDISC had earlier released six earlier documents on therapeutic area standards [28], the products had generally been limited in scope and often took two or more years to complete. CFAST provided a vehicle for TCB to contribute financial and dedicated in-kind resources to help speed up the development of new standards under an enhanced CDISC process that previously depended largely on the availability of volunteers. The first new standards released under the CFAST initiative were completed in less than a year, with products that addressed a wider scope that was intended to be meaningful to clinical and medical experts, as well as data managers and statistical programmers.

While these therapeutic area standards provided useful insight into information elements that might be associated with disease areas, the success of the effort was limited because of a decision to focus only on pharmaceutical research requirements rather than trying to capitalize on the vast amount of information that can be captured in academic research, primary healthcare scenarios and standard of care treatment plans. So, while the CFAST standards had the potential to be more directly relevant to research and application of healthcare protocols and practices, the projects persisted in defining new terminologies rather than adopting the existing terminologies (e.g., LOINC, SNOMED CT) that are better understood and already in wide use across the much more expansive healthcare industry.

Realizing the promise—putting standards to work

While evolving regulatory requirements and increased industry acceptance of data standards eventually opened the door to finally achieving some of the benefits long promised by data standards, it is clear that much of the real work is still to come. As more and more therapeutic area standards are produced (and a number of new CFAST standards have been produced with each successive year), the burden of ensuring consistency and reusability increases, so the prior paradigm of getting people together to work with spreadsheets and documents over telephone meetings was recognized as no longer sufficient to support the expanding need. It was also clear that standards had to walk a thin tightrope between both ensuring that early adopters do not have to change course midstream as new versions of standards are issued, while also being able to more rapidly produce new works—even as science continues to evolve with new methods and areas of interest that require the standards to grow and adapt.

The need to provide a systems infrastructure to develop, house, and deliver standards as electronic metadata led to the development of the CDISC Shared Health and Clinical Research Electronic Library (SHARE) [29]. CDISC SHARE was intended to accelerate the development of new therapeutic standards based on reusable high-level research conceptual components, and deliver standard metadata directly to systems for implementation across the full project life cycle from design through analysis. SHARE also aspired to make it possible to express relationships not just between the various CDISC standards and concepts, but also externally, to potentially provide a platform to integrate public ontologies and concepts used in the conduct of healthcare. In 2019, the SHARE platform was upgraded to a new technology platform and renamed to the CDISC Library [30]. Of course, having a standard metadata repository became essential because of the many variations among CDISC standards, since there were incompatibilities between many of the individual CDISC standards. While the Library provides one-stop shopping for the full spectrum of standards, it is still necessary to conduct transformations by mapping between source systems and CDISC standards, and indeed from one CDISC standard to another. While the SHARE/Library tool makes it easier for computers to process these conversions, it is also a time-consuming task still requiring manual intervention with a significant risk of introducing error in each step. A more elegant solution would be to minimize mapping transformations across the research process, ideally by utilizing a common set of standards that can support primary healthcare use cases as well as research.

The potential of reusing healthcare information for research purposes and embedding research concepts more in the conduct of healthcare-related services to many more patients has long offered a compelling vision for vastly improving the health and care of patients in the future. The Institute of Medicine, for example, explored this intriguing idea in their report on the Learning Healthcare System [31].

But translating the vision of linking healthcare and research through semantic interoperability into reality has been an imposing challenge [32,33]. An FDA-sponsored attempt to create a message-based healthcare standard consistent with the CDISC CDASH and SDTM standards based on the Health Level Seven (HL7) [34] version 3 Reference Information Model (RIM) in the mid-2000s [35] proved unwieldy and overly complex—and was thus never adopted [36]. And, for those involved in pharmaceutical clinical research data, an ongoing topic of debate asked whether research data are really the same as healthcare data [37]. The contention was that research studies are protocol-driven, and these experimental protocols precisely define how to treat the patient consistent with the experimental design—which may vary considerably from standards of care—as well as how to collect the data. Research data are historically presented in tabular files of rows and columns, which allow aggregation, sorting, and tabulation of results. Healthcare data is directed toward patient care, not research, and has often been considered to be of dubious quality. Such clinical data does not typically include many of the detailed CRF questions that accompany a study to measure safety and efficacy. Still, it was acknowledged that there is overlap between these two data spheres: such as basic demographics, concomitant medications, and observed data such as vital signs and especially laboratory data. It is also clear there is overlap in medical history although medical history questionnaires tend to be very diverse, tailored for specific cases (rather than a comprehensive, standardized patient history which should be part of the personal health record that a patient would present to their physician). And, historically, most healthcare clinical data, was more likely to be captured as scanned documents or patient notes rather than structured, consistent data. Thus, reuse of healthcare data was not so easy— protocol driven research collects data differently than the more reactive world of primary healthcare, and the systems are not aligned (having different business, scientific and especially regional regulatory requirements and attitudes). Furthermore, as noted above, research typically uses different semantic vocabularies for coding information (for example, healthcare typically codes problems and diagnoses in ICD10 or SNOMED, whereas clinical research records adverse events in MedDRA).

Nevertheless, CDISC explored this topic by collaborating with Integrating the Healthcare Enterprise (IHE) on a set of profiles [38] that enable the use of Electronic Healthcare Record data for multiple research purposes, including patient recruitment and CRF data capture [39]. In Europe, the Innovative Medicines Initiative (IMI) funded the EHR4CR project [40] applying similarim methods, and other efforts were also explored in Japan and other countries. Some proof-of-concept projects applied these methods, but these were not deployed on a large scale because of system integration and scalability challenges, including the necessity of adding still more mapping transformation steps for anyone who tried.

A path forward with healthcare standards for real-world data

So, the use of the HL7 v2 and v3 health standards for research never achieved much traction, while CDISC standards were dedicated only to research, with no practical application to other healthcare use cases. But this attitude began to change with the arrival of HL7's Fast Healthcare Interoperability Resources (FHIR) in 2011 [41]. HL7 FHIR was conceived as a next generation platform standard, based on modern web technologies and a standard data model built upon resources, designed explicitly to support all types of use cases for healthcare—including research. This was transformational, because FHIR offered a common API that could represent healthcare data from EHR systems in a consistent manner. Use of FHIR has been mandated in the United States, [42] is becoming a mainstay in other countries and has stimulated a flurry of activity worldwide. FHIR now offers an opportunity to truly realize the vision of the Learning Health System—with a common standard that can support research as well as the rest of healthcare. FHIR's use to support value-based care [43] in the USA has also increased the breadth and quality of healthcare data for research. The benefits of the FHIR standard to support needs to use Real-World Evidence and research has been recognized by a group of former FDA Commissioners [44] and the US National Institutes of Health [45] and many other governmental as well as other healthcare industry bodies across the globe. Use of HL7 FHIR presents a previously unprecedented opportunity to effectively converge healthcare and research. While CDISC provides solutions for today's world, FHIR provides a path forward to a future built on interoperability, in some cases potentially replacing certain CDISC standards (such as SDTM and CDASH) while converging with others (such as ODM and ADaM).

SWOT analysis

While the many benefits and opportunities of adopting clinical data standards are compelling, as indicated above, progress has been slow due to several perceived weaknesses and threats:

Strengths

- Data standards increase productivity in study setup and analysis.
- Standards make it possible to more rapidly glean knowledge from data and promote data reusability.
- Use of common metadata representation standards such as the HL7 FHIR StructureDefinition, content standards such as Patient and Observation, and protocol-related standards such as PlanDefinition and CarePlan make it possible to impose more control on studies that can also be directly interpreted and applied by EHR systems.

Weaknesses

- Adopting standards requires investment in adapting systems and processes which companies may be reluctant to commit.
- Data standards are presumed by some researchers to pose constraints on the scientific creativity, even though others believe they actually make it possible to advance science.

Opportunities

- Many regulatory authorities are now encouraging or requiring use of data standards.
- Adoption of a common, uniform set of data standards can promote precompetitive cooperation among research participants.
- A demand for real-world evidence suggests a reexamination of whether research outside of healthcare is still necessary or desirable compared to research capabilities fully integrated with healthcare.
- A more complete solution toward interoperability could also involve common healthcare technologies rather than those limited to pharma research only (SNOMED vs. MedDRA) and more advanced semantics such as detailed logical models [46].

Threats

- Unclear regulatory statements about which standards to use have led some sponsors to delay investing in standards adoption, in the fear that the standards they adopt now may differ from what is ultimately required.
- Some sponsors and CROs may still fear they may contribute to unintentional disclosure of intellectual property or expose potential product weaknesses to competitors.

Take-home message

Just as any effective communication between two parties is based on use of a common language, so do clinical data standards provide the common language for fully realizing the extraordinary opportunity of more efficient clinical research, through improved productivity, increased knowledge, more effective dialogue with regulatory authorities, and expanded collaborations throughout the global research community. Much progress has been achieved to date largely through the efforts of CDISC, but it appears we are on the cusp of transforming research by making research more interoperable with healthcare, ideally using a common set of modern, common platform standards such as HL7 FHIR in the future. The best is yet to come.

Table 16.1 an overview of CDISC data standards for clinical research (as of 2016).

Table 16.1 Overview of principle CDISC standards for clinical research.

Research application area	CDISC standard	Description	Current version/date	Other relevant standards
Foundational				
Study Planning	Protocol Representation Model (PRM)	BRIDG-based model representing standard protocol content elements and relationships	V1 2010	Protocol Toolset
Data Collection/Study Conduct	CDASH	Clinical Data Acquisitions Standards Harmonization – describes basic data collection fields for CRF data with guidelines, best practices	V2.0 2017	CDASH Model, CDASH IG
Data Collection/Study Conduct	CDASH SAE Supplement	CDASH standard describing basic data collection fields for ICH E2B SAE data with implementation guidelines, best practices	V1 2013	ICH E2B HL7/ISO ICSR
Data Collection/Study Conduct	LAB	Standard model for the acquisition and interchange of clinical lab data between lab and sponsor/CRO recipients	V1.0.1 2003	HL7 v2 Lab Message LOINC
Submission/Analysis	ADaM	Analysis Data Model – describes fundamental principles and standards for creating analysis datasets and metadata	V2.1 2009[a]	None
Submission/Analysis	ADaM IG	Implementation Guide that describes standard data structures, conventions and variables used with the ADaM model	V1.2 2019	Incorporates multiple documents.
Foundational (SDTM Family)				
Submission/Tabulation	SDTM	Study Data Tabulation Model describes principles of representing clinical and non-clinical tabulation data	V1.8 2019	None
Submission/Tabulation	SDTMIG	SDTM Implementation Guide for Human Clinical Trials (Drug Products and Biologics)	V3.3 2018	

Submission/Tabulation	SEND	Standard for Exchange of Nonclinical Data: SDTM IG to represent data from nonclinical studies	V3.1 2016[a]	SEND DART, CoDEx. Animal Rule
Submission/Tabulation	SDTMIG-MD	SDTM Implementation Guide for Medical Devices — to represent data from clinical trials using medical devices	V1.1 2019	
Foundational (XML Data Exchange)				
Study Planning	Study Design Model (SDM-XML)	XML schema specification based on ODM for representing clinical study design, including structure, workflow and timing	V1 2011	HL7 Study Design Structured Document (DSTU)
Data Collection/Study Conduct	ODM	CDISC standard for the regulatory compliant acquisition, exchange and archive of clinical trials data and metadata	V1.3.2 2013	None
Submission Metadata	Define-XML	XML schema specification to describe metadata for SDTM, SEND, and ADaM submission datasets	V2.1 2019[a]	None
Submission/Dataset Exchange	Dataset-XML	XML schema specification for representing study datasets associated with Define-XML metadata	V1 2014	SAS V5 Transport
Semantics				
Semantics	Glossary	Glossary with definitions of acronyms and terms commonly used in clinical research	V13 2019	None
Semantics	Controlled Terminology	Controlled terminology to support CDISC standards such as SDTM, CDASH, ADaM (in partnership with NCI EVS)	Quarterly; Pkg 40 2019	MedDRA, SNOMED CT …
Semantics/Model	BRIDG	Biomedical Research Integrated Domain Group – UML model of the semantics of protocol-driven clinical research	V5 2017	None
Semantics/Metadata	CDISC Library/SHARE	CDISC Metadata Repository – electronic source for all CDISC standard metadata and terminology	2019 Release	NCI CaDSR
Implementations				
Study Planning > Submission/Analysis	Therapeutic Area Standards	User Guides describing how to apply CDISC Foundational Standards with controlled terminologies on clinical studies in specific disease areas	Ongoing since 2011	Various Ontology and Common Data Element libraries

Continued

Table 16.1 Overview of principle CDISC standards for clinical research.—*cont'd*

Research application area	CDISC standard	Description	Current version/ date	Other relevant standards
Study Planning > Submission/ Tabulation	Questionnaires	SDTM Implementation Guide Supplements with annotated CRFs and Controlled Terminology for representing data from Questionnaires commonly used in clinical studies	Ongoing since 2013	Structured Data Capture
Study Planning > Conduct	Healthcare Link	A suite of CDISC and IHE standards and 'enablers' to improve the workflow of clinicians doing research leveraging electronic health records and research systems	Project initiated in 2004	Rebranded as Real World Data

[a] *Current or prior versions of Specification in FDA Guidance.*
Source: https://www.cdisc.org/standards — used by permission.

References

[1] CDISC website. www.cdisc.org. [Accessed January 2020].

[2] FDA Press Release. Available at: http://www.fda.gov/newsevents/newsroom/pressannouncements/2004/ucm108330.htm. [Accessed January 2014].

[3] v7/22. FDA Study Data Standards Page; 2020. [Accessed 30 Aug 2020].

[4] Rozwell C, Kush RD, Helton E, Newby F, Mason T. The business case for CDISC standards. Gartner; 22 March 2006. Web, www.gartner.com/doc/490510. [Accessed January 2020].

[5] CDISC study data tabulation model implementation guide: human clinical trials, Version 3.2. Available at: www.cdisc.org/sdtm. [Accessed January 2020].

[6] Willoughby C, Fridsma D, Chatterjee L, Speakman J, Evans J, Kush R. A standard computable clinical trial protocol: the role of the BRIDG model. Drug Inf J 2007;41: 383—92.

[7] CDISC Clinical Data Acquisition Standards Harmonization (CDASH) version 1.1. Available at: https://www.cdisc.org/standards/foundational/cdash. [Accessed January 2020].

[8] Kubick, W.R., Buckthorn wars and the essence of clinical data. Available at: http://www.appliedclinicaltrialsonline.com/appliedclinicaltrials/article/articleDetail.jsp?id=776909. [Accessed January 2020].

[9] Kubick WR. The elegant machine: applying technology to optimize clinical trials. Drug Inf J 1998;32:861—9.

[10] Kubick WR, Ruberg S, Helton E. Toward a comprehensive CDISC submission data standard. Drug Inf J 2007;41:373—82.

[11] Kush RD, Bleicher P, Kubick W, Kush ST, Marks R, Raymond S, Tardiff B. eClinical trials: planning and Implementation. Boston: Thompson Center Watch; 2003.

[12] FDA study data standards resources. Available at: https://www.fda.gov/industry/fda-resources-data-standards/study-data-standards-resources. [Accessed January 2020].

[13] The critical path initiative. Available at: http://www.fda.gov/downloads/scienceresearch/specialtopics/criticalpathinitiative/ucm221651.pdf. [Accessed January 2014].

[14] CBER/CDER study data standards for regulatory submissions position statement. Available at: https://www.fda.gov/industry/study-data-standards-resources/study-data-submission-cder-and-cber. [Accessed January 2020].

[15] European Medicines Agency draft policy 70: publication and access to clinical-trial da20. http://www.ema.europa.eu/ema/index.jsp?curl=pages/includes/document/document_detail.jsp?webContentId=WC500144730&mid=WC0b01ac058009a3dc. [Accessed January 2014].

[16] PMDA request for electronic clinical study data for pilot project. Available at: http://www.pmda.go.jp/operations/shonin/info/iyaku/jisedai/file/tsuuchi_e.pdf. [Accessed January 2014].

[17] Presentation by norman stockbridge at DIA/FDA CDER/CBER computational science annual meeting march 22—23. 2010. http://www.diahome.org/Tools/Content.aspx?type=eopdf&file=%2fproductfiles%2f21575%2f10014%2Epdf. [Accessed January 2014].

[18] Kubick, W.R., When great ideas meet the real world. Available at: http://www.appliedclinicaltrialsonline.com/appliedclinicaltrials/CRO%2FSponsor/When-Great-Ideas-Meet-the-Real-World/ArticleStandard/Article/detail/655353?contextCategoryId=554. [Accessed January 2020].

[19] Janus Clinical Trials Repository (CTR) project. http://www.fda.gov/forindustry/datastandards/studydatastandards/ucm155327.htm. [Accessed January 2014].

[20] Kubick, W.R., The power and pitfalls of aggregate data, Available at: http://www.appliedclinicaltrialsonline.com/appliedclinicaltrials/IT/The-Power-and-Pitfalls-of-Aggregate-Data/ArticleStandard/Article/detail/742590?contextCategoryId=554. [Accessed January 2014].

[21] Haber M, Kisler BW, Lenzen M, Wright LW. Controlled terminology for clinical research: a collaboration between CDISC and NCI enterprise vocabulary services. Drug Inf J 2007;41:405—12.

[22] Goldman M. New frontiers for collaborative research. Sci Transl Med 2013;5:216ed22. https://doi.org/10.1126/scitranslmed.3007990.

[23] TCB press release or website. http://www.prnewswire.co.uk/news-releases/ten-pharmaceutical-companies-unite-to-accelerate-development-of-new-medicines-170329496.html. [Accessed January 2020].

[24] TCB initiatives website. https://transceleratebiopharmainc.com/what-we-do/. [Accessed January 2020].

[25] PDUFA reauthorization performance goals and procedures fiscal years 2013 through 2017. http://www.fda.gov/downloads/forindustry/userfees/prescriptiondruguserfee/ucm270412.pdf. p. 28. [Accessed January 2020].

[26] C-path website. http://c-path.org. [Accessed January 2014].

[27] CFAST press release. CDISC, C-path, FDA, TransCelerate and the global CDISC community launch initiative to accelerate therapies through standards. Available at: http://www.cdisc.org/cfast. [Accessed January 2020].

[28] CDISC therapeutic area standards webpage. https://www.cdisc.org/standards/therapeutic-areas. [Accessed January 2020].

[29] CDISC SHARE/library archives webpage. https://www.cdisc.org/cdisc-library-archives. [Accessed January 2020].

[30] CDISC library launch webpage. https://www.cdisc.org/events/education/webinars/2019/02/cdisc-library-launch.

[31] Institute of medicine. Digital infrastructure for the learning health system: the foundation for continuous improvement in health and health care — workshop series summary. Available at: http://www.nationalacademies.org/hmd/~/media/Files/Report%20Files/2010/Digital-Infrastructure-for-the-Learning-Health-System/Digital%20Infrastructure%20Workshop%20Highlights.pdf. [Accessed January 2020].

[32] Mead C. Data interchange standards in healthcare IT — computable semantic interoperability; now possible but still difficult, do we really need a better mousetrap? J Healthc Inf Manag 2006;20:71—8.

[33] Kush RD, Helton E, Rockhold FW, Hardison CD. Electronic health records, medical research and the Tower of Babel. N Engl J Med 2008;358:1738—40.

[34] HL7 website. www.hl7.org. [Accessed January 2020].

[35] FDA webpage. https://www.fda.gov/media/76855/download.

[36] Norman, K., Simplifying data standards: CDISC and HL7 deciphered. Available at: http://www.appliedclinicaltrialsonline.com/appliedclinicaltrials/article/articleDetail.jsp?id=672337. [Last accessed January 2020].

[37] Kush RD. Data sharing: electronic health records and research interoperability. In: Richesson R, Andrews J, editors. Clinical research informatics. Springer Publ.; 2012. p. 313—34.

[38] Integrating the Healthcare Enterprise (IHE) website. http://www.ihe.net.

[39] IHE healthcare link profiles Available at: http://www.cdisc.org/stuff/contentmgr/files/0/
f5a0121d251a348a87466028e156d3c3/misc/cdisc_healthcare_link_profiles.pdf.
[Accessed January 2020].

[40] EHR4CR website. http://www.ehr4cr.eu, https://www.cdisc.org/news/fda-binding-
guidance-goes-effect-december-17th. [Accessed January 2020].

[41] HL7 website. http://hl7.org/fhir/.

[42] https://www.healthdatamanagement.com/news/proposed-onc-rule-requires-fhir-
interoperability-standard.

[43] A prime example is the HL7 Da Vinci program. http://www.hl7.org/about/davinci/.

[44] https://bipartisanpolicy.org/report/expanding-the-use-of-realworld-evidence-in-
regulatory-and-value-based-payment-decision-making-for-drugs-and-biologics/.

[45] https://datascience.nih.gov/fhir-initiatives.

[46] https://www.logicahealth.org/about/roadmap/.

Use of data analytics for remote monitoring

17

Gareth Milborrow

Doctor, SVP Data & Applied Analytics IT ICON Clinical Research, Southampton, United Kingdom

The need

The fundamental motivation for clinical data analytics is to improve quality in clinical trials without the need for excessive on-site activities. We define quality as "the absence of errors that matter in decision making for patients." How do we move from processing and analyzing data to a place where we can make rational and consistent decisions to improve quality? First, we have to eliminate the data that does not matter in order to start to make sense of the data that does matter. Traditionally, attempts to reduce the collection of extraneous data were aimed at achieving this goal. Although this did enrich the data, it did not come close to achieving the critical mass necessary to ignite a chain reaction to convert the data into useful information (let alone knowledge). Much of the data does matter but only when aligned and compared in context and in such a manner as to transverse the divide so that information can then be extracted (or visualized). When visualization is then coupled with effective and timely decision making, you have a situation where the power of the data is fully harnessed, and the decision-making process becomes almost intuitive for those able to "read" the knowledge stream.

The solution

Staring at a spreadsheet of uniformly gray digits on your computer screen, you would be forgiven for being under the impression that the data represents a uniformly objective truth that is static in nature and closely, if not absolutely, tied to the site. On the surface, there is very little to distinguish the history of the data on one tab from another, giving the illusion that all data has a common history and that all data came to appear on your screen through a uniform process that affected one dataset no more than any other; an impression that leads you to believe that all data holds more or less equal value (or information) in the detection of risk.

This could not be further from the truth.

Not all data is what it seems, and not all data holds the unbiased, transparent value we often hope it should. Instead, the detection of risk through the analysis of data requires a detailed understanding of the origin of the data and the path it took before it landed on your computer screen. Failure to fully appreciate the dynamic nature of the data, its life, and its purity will almost certainly undermine the value of the subsequent analysis.

Let's start with an apparently simple example: vital sign data. Typically, we are presented with three core columns containing the systolic measurement, the diastolic measurement, and the heart rate measurement, all taken simultaneously at some point in history. On the surface, this is a relatively simple dataset; a patient identifier followed by a series of numbers (the vital sign data) alongside the metadata describing the time and date the data was taken (and possibly some other clinical parameters, for example, the patient's position).

But, in reality, this dataset is far from straightforward. For a start, some of those measurements were taken on digital blood pressure meters and some were performed manually. Furthermore, the percentage of readings taken on digital meters varies from site to site, country to country, and study to study based on therapeutic indication (cardiology studies have a particularly high percentage of readings taken on digital meters), regional wealth (Western countries tend to have more digital readings than third world or emerging markets due the expense of the meters), and personal preference of the site staff This is important. Analysis of the data derived from digital meters reflects the accuracy of the meter, not the integrity of the site. However, data obtained from a manual measurement, in contrast to a digital measurement, is permanently and very closely linked to the site's integrity for several reasons listed in the following text.

First, it is difficult for anybody to retrospectively influence the data. To better understand this point, let's briefly examine another dataset-concomitant medication (conmeds). In the author's experience, it is not unusual for a site to omit a number of conmeds from the patient's case report form (CRF), probably due to the burden on the site and because of a perceived lack of value of the data. However, even if the site does not document all of the conmeds, a vigilant clinical research associate (CRA) will detect the omissions and request that the conmeds are added to the CRF. It is vitally important that you understand this subtle event. Initially, the site did not perform to the standards required, but following the actions of the CRA, the site added the missing data. If a risk analysis was performed on the conmed data prior to the CRA visit, there was a good chance we would have detected the site's inappropriate behavior, but once the CRA took corrective action, analysis of the conmed data would [probably] no longer detect the noncompliance. This transition reflects the subtle but important difference between quality and risk. Following the CRA's corrective action, the quality of the data improved (particularly its accuracy and completeness) but the risks may not have changed. Risk is a trait, not a data point. Critically, our subsequent ability to detect that risk trait has been masked by the corrective action. As a result, the subsequent value of the conmed data as an indicator of risk is greatly reduced because we are no longer analyzing the activities of the site but, instead, the collective actions of the site, the CRA, and the site's compliance with the corrective actions—statistically referred to as confounding variables. Trace evidence tends to remain, but it can be heavily masked. We will return to conmeds later in this section, but for now, let's get back to the vitals data.

Unlike conmeds, it is very difficult, if not impossible, for the CRA to retrospectively influence the vitals data. Source data verification (SDV) merely serves to confirm that the data in the CRF is consistent with the data in the source documents. Even if the CRA became suspicious that a specific blood pressure reading of 120/80 mm Hg appeared improbable, considering the sea of blood pressure data with identical values, it is very difficult for the CRA to retrospectively argue the point. After all, the CRA was not there on the day of the measurement, and who is to say the reading was not 120/80? Or so the investigator will argue. Also, the CRA does not have a global, statistically supported view of the data. Instead, they are embedded in the details of the data, reviewing each individual data point and compiling a limited impression of the site. Therefore, aside from making corrections to transcription errors, vital sign data goes largely unaltered throughout the life cycle of study. To be clear, this is not to say that the views and actions of the CRA are irrelevant, but the message is that the CRA can only operate within the confines of human capability and the process.

A second valuable feature of vital sign data derived from manual measurements is that the manual blood pressure readings are strongly influenced by the operator (digital readings are not of course).This may seem obvious, but the implications are profound, and the ability to detect these influences are not always immediately obvious to an analyst without a medical background. For example, a typical approach to the analysis of blood pressure data when in pursuit of risk (or fraud) detection is to calculate the standard deviation (or variance) of the systolic or diastolic measurements and to compare the result on a per site basis with the other sites collectively. If the standard deviation is unusually high or low at any one site, the conclusion is that there maybe a problem, and a CRA should be dispatched. This approach is fraught with problems, not least of which is the question "What do we expect the CRA to do about it?" But, let's look at some characteristics of the data first.

Systolic and diastolic measurements are highly correlated. As the systolic reading increases, so does the diastolic reading (typically by about half of the systolic increase). This correlation is ingrained in the clinician over the years of his or her practice but, importantly, the correlation is by no means universal or absolute. In some diseases, for example, aortic regurgitation, the exact opposite may occur, and the diastolic reading may, in fact, decrease as the systolic increases, albeit it is a rarity by comparison.

Number preference is widely used: clinicians often round off the blood pressure readings to the nearest five or zero, for example, 120/80 or 135/85, and this is generally considered acceptable in routine clinical practice. Digital blood pressure meters do not do this.

Last, the range of available readings is often surprisingly narrow. In a study requiring normotensive patients, the acceptable range of systolic measurements might be between 100 and 140 mm Hg. But if the clinician is rounding off to the nearest 5, this is a mere nine systolic values at baseline (100, 105, 110, 115, 120, 125, 130, 135, and 140 mm Hg systolic) and far less if there are only one or two patients who will each remain within a biologically narrow systolic range over time.

With this in mind, the use of standard deviation as a tool for detection of risk using blood pressure data becomes challenging. Standard deviation can only be used on either systolic or diastolic measurements independently (there are exceptions, but they have their own limitations), does not take into account the difference between manual and electronic blood pressure readings, and is a measure of data spread in a data range which is often inherently narrower than anticipated. That is not to say it does not work. In fact, it does work, but only to a degree. In the author's experience, standard deviation worked well to detect those sites that had unrelentingly repeated the measurement 120/80 over and over, but it did not work well when the investigator deliberately attempted to deceive—arguably a site of even greater concern. In cases where there is an attempt to deceive, the investigator may deliberately spread the data over a few systolic and diastolic readings, for example, 140/90, 120/80, and possibly 110/70. However, because there are only nine possible systolic values available to the investigator (presuming he/she rounds off to 5 mm Hg), as compared to the 41 possible values available to a digital meter, it is very easy for him/her to create the illusion of adequate data spread and avoid detection, often deflecting our attention to smaller sites with few patients and, thus, a very limited range of systolic data, albeit completely genuine data.

Instead of standard deviation, what is needed is an algorithm(s) which takes into account the existence of digital blood pressure meters, the correlation of systolic and diastolic data, the intra- and interpatient variability, the presence of number preference, and the interoperator variability, ideally presented to the user in a simple to understand format. The key to the algorithm(s) is to capitalize on the [fraudulent] investigator's ingrained sense of correlation and staccato data spread, neither of which replicate real data.

When this is achieved and combined with a well-defined stepwise approach to the analysis, the blood pressure readings, unlike conmeds, provide a persistently valuable window into the site's risk profile that is largely untainted by repeated monitoring and that is continuously updated throughout the life cycle of the site as new readings are taken (the high frequency of vital sign measurements is another valuable benefit of this dataset).

Like blood pressure data, the heart rate data is also more complicated than initially meets the eye because heart rate data is often extrapolated, not actual. As a result, it is our experience that standard deviation does not work as well as we hope and, like blood pressure, we need to create an algorithm(s) which is specifically suited to heart rate data to detect those sites that are deliberately and systematically fabricating data. In our experience, heart rate data does have one small but significant advantage over blood pressure data: using a good deal of clinical acumen, it is possible to use simple algorithms to instantly demonstrate that the data is fabricated. When combined with evidence of suspicious blood pressure data, the verdict is almost conclusive, and we are able to direct the CRA to conduct very specific, well-defined procedures to prove it. This is the advantage of a well-thought out, carefully executed risk detection and management plan.

A well-constructed risk detection platform is one which provides sensitive and specific risk-detection algorithms that cover the study over its whole life cycle, cover all the roles in the site, are unaffected by monitoring.

On-site activities continue to play a key role in quality management.

These activities include confirmation of subject-informed consent; investigational product accountability; site file review; site relationship management; source data review; and source data verification (particularly of critical data).

However, on-site monitoring alone is not well suited to the detection of comparative performance signals, trends, and patterns across multiple subjects, sites, and countries, or the monitoring of uncommon events to identify sites that may need corrective action. These activities are best performed by centralized monitoring of aggregated data using technology specifically designed for these purposes.

The first stage of process is to determine the specific risks and global risk-level associated with the clinical trial and protocol, in line with quality by design principles. Optimally, this process begins during early development of the protocol prior to protocol finalization. The project team meets with the relevant stakeholders to discuss and define these risks. Multiple criteria relating to the trial protocol, patient population, investigational product, data management, and geographic criteria are evaluated. A composite study risk score is calculated for each criterion. The risk score informs the cross-functional team's discussion of the integrated cross-functional risk management strategy.

A risk management plan is developed using the outputs of the risk assessment process. This document defines plans for a range of activities, including communication, governance, quality assurance, medical monitoring, data management, study start up, vendor management, documentation control, and site monitoring (Fig. 17.1).

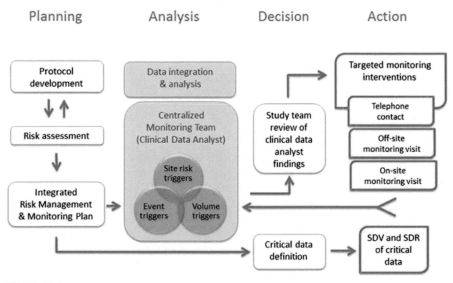

FIGURE 17.1

Overview of an integrated remote/on-site monitoring process.

It also defines the risk indicators that are analyzed by the centralized (remote) monitoring team to identify variant or outlier site performance and the Quality Tolerance Limits that will define outlier sites and trigger an intervention (Fig. 17.2).

(A)

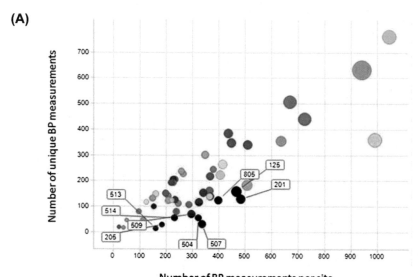

(B)

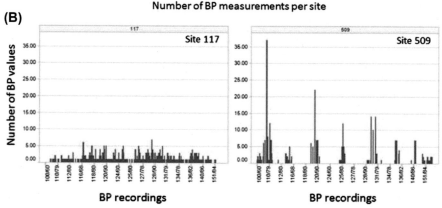

FIGURE 17.2

The Quality Risk Indicator Blood Pressure. (A) shows the distribution of sites, comparing the total number of unique BP data points reported at each site. A low such number indicates that certain BP values are reported multiple times. Highlighted sites (black) are the sites with the lowest 15% of the distribution. (B) shows data distribution histograms from two sites: Site 117 is a typical "quality" site, reporting 366 individual blood pressure (BP) measurements from 13 subjects, with the BP values distributed across a wide range. In contrast, site 506, which is one of the outlier sites from (A), has only 288 values reported from 11 subjects, distributed in distinct groups, with only 70 unique recordings. This suggests the site is rounding BP values to the next nearest 10 mmHg.

Such risk indicator triggers override data volume triggers, which override milestone triggers:

- Risk indicator triggers: Comparative evaluation of site activity across the study, according to objective risk indicators, indicates the need for the site contact to perform investigative or corrective action. Risk triggers are considered in three domains: "recruitment" (e.g., screening to randomized ratios, withdrawal rates), "reporting diligence" (e.g., adverse events reported per randomized subject per month exposure; concomitant medication per randomized subject), and "data quality" (e.g., variability index of vital sign data, to identify digit preference, data rounding tendencies, or biologically implausible data).
- Data volume triggers: For example, a site contact is recommended when the volume of data or subjects at a site are above a defined threshold to justify a full day on-site visit to perform SDV, SDR, and other on-site activities.
- Milestone triggers: These are triggers relating to defined milestones at a site that initiate an on-site monitoring visit, for example, the first subject randomized at a site.

The monitoring document defines the recommended monitoring interventions based on the significance or severity of the finding.

The following example of visualization relates to the initially outlined example of vital signs as a risk indicator for data quality. Using a suit of dynamic filters and drill-down functions, the remote monitor can interact with the data to generate further analytical outputs (Fig. 17.2).

Remote monitoring employs additional site monitoring activities to ensure all site performance and compliance issues, including remote monitoring findings, are addressed through on-site monitoring visits and telephone contact.

Any additional monitoring interventions that are decided and performed to address CDA findings by the CRA are documented either in a Clinical Trial Management System (CTMS) or—if through telephone contact— by use of a telephone contact record form that is filed as part of the trial master file (TMF).

Such remote monitoring covers all core aspects necessary to perform interactive monitoring of clinical sites and their performance.

Take-home message

The aim of centralised analytics is to improve the safety of the patient — both in the current trial and future patients who may use the approved therapy — and to increase the likelihood of a definitive outcome to the study by improving the overall quality of the study execution and the data itself.

Further reading

[1] Guidance for Industry. Oversight of clinical investigations a risk-based approach to monitoring. 2013. www.fda.govd/ownloads/drugs/guidancecomplianceregulatoryinfor mation/guidances/ucm269919.pdf.

[2] Reflection paper on risk based quality management in clinical trials. 2013. http://www.ema.europa.eu/docs/en_GB/document_library/Scientific_guideline/2013/11/WCS00155 491.pdf.

[3] Position paper: risk-based monitoring methodology. 2013. http://www.transceleratebiopharmainc.com/content/risk-based-monitoring-methodology-position-paper/.

Taking control of high data volumes

18

Henrik Nakskov

Director Management CIMS 2015 Aps, Charlottenlund, Danmark

The need

Unlike a century ago, when economic growth was dependent on factors such as good supply chain management, the industrial assembly line, and exchange of physical goods, today's modern economic growth is dependent on the ability to handle and manage large volumes of often unstructured information.

The Corona pandemic massively accelerated that need. The variety of data sources in clinical trials is rapidly growing, not only including EDC data but also information from e-diaries, wearables, smartphones, health records, and eventually also from social media. The competitive advantages of good process methodology and well-trained staff have been replaced by the ability to structure information and react fast.

Modern organizations rely on structured information and data intelligence to make decisions. However, all the various forms of data intelligence structures, such as reports, visuals, listings, and other indicators have one fundamental flaw that is often overlooked: Human nature can never be expected to react predictably. Although the data might be neatly arranged in a predefined structure, indicating predictability and certainty, when humans are involved, discrepancies too become involved.

When we consider that the data shown in these various structures is the result of human activity, it becomes clear that regardless of the sophisticated technology involved, blind reliance on data is not the way forward. Critical thought and the ability to question data still remain highly relevant. In the field of clinical drug development, this is even truer as problems with data quality can have far-reaching consequences that can ultimately affect patient safety and an organization's very existence.

The movement of data through the clinical trial process can in many ways be compared to the shipping of valuable content. The ship in itself is not important; what is important is that it has the required stability to transport the goods, is able to dock at the harbor, and that it can unload and load its valuable content safely and efficiently. In the same way, we have to maintain focus on the data that is being transported and moved. It is just an enabler.

There is a considerable risk of poor sciences associated with fragile handling of data and false interpretations of data. This leads to false causality and the consequences can be medication being developed which did not reach its full potential or are excluding patients' segments that could have been treated for a medical condition. From a data integrity perspective, decisions based upon poor casualty quality data are in best case worth nothing.

The 3T method and perspective should be applied before making any foundational decision, because only risk-based thinking ensures health and safety for the patients. It can be used as a remarkably simple tool which gives a measurement of the level of data integrity. The robustness of the decisions based on such data is proportional to the integrity.

- Truth: Is data a reflection of the occurrences and events under which it was captured? Does data tell the truth?
- Trust: When we handle data, do we then ensure that any change, conversion, and interpretation can be trusted? Is our activity with data validated and verifiable?
- Traceability: Can we trace data both upstream and downstream without any ambiguity. Is the line of data handling from origin to interpretation clear and free of ambiguity?

The level of data integrity can very easily be calculated. It's simply 3*the level of integrity of the 3T: If all has a level of 80%, then we only have 51% overall integrity—what comes close to tossing a coin.

The solution

This chapter will look at ways to help you navigate through complex clinical trial data and information structures without drowning in complexity and technology. In particular, we will look at:

- Using the growing wave of data to your advantage.
- How managing data complexity is not just about technology.
- How processes, organizational psychology, and people need to learn to use technology and data to their best advantage.
- How to use past and real-time data to improve and develop.
- Combined, all of these points will lead to the foundation of future decision making, knowledge leadership, and knowledge management. For details of knowledge management also see next section "Share the (digital) knowledge based on quality data" by L Hyveled.

Taking control over the seemingly uncontrollable world of data is actually very straightforward and can be described in eight stages. By mastering all the eight stages, you will be on the way to mastering data quality. However, before we look at the eight stages, it is necessary to understand the competencies that are needed.

The ability to be successful as an information supply chain unit requires business context and process knowledge. Ideally, you should possess a high degree of both in-depth specialized knowledge and an understanding of all the interconnected processes. These skills involve being able to talk to people, investigate connections, draw process diagrams, handle data exchange connections, evaluate the impact of influences, know the system's data conversion, create communication channels, and finally, have the ability to "sell" your information.

Although you may not consider yourself to be a sales person, being able to create buy-in and acceptance from the business owners and stakeholders you work with will decrease reluctance to accept change and hasten the progress toward the goal of better data quality. By getting to know the human element of your business, you also have a better chance of predicting the unpredictability of your customers and preempting their needs in the future.

Stage 1 understand the context

Data management is the development, execution, and supervision of plans, policies, programs, and practices that control, protect, deliver, and enhance the value of data and information assets. In short, it is about handling the fastest growing part of our world. In the clinical process, data comes from two main sources: clinical trial data from patients and clinical trial conduct data related to the actual execution of the trial.

The potential and competitive advantage of data comes from the ability to control the flow of changes that originate from the data-information-knowledge growth. This is not just about technology. It is about understanding data and interpreting data in the context of its origin. Good information management is about data causality. In other words, it is about the ability to understand the connections that come from nonlinear volumes of information. By being able to understand the business context and the clinical process and then applying this understanding to the data, you can see cause and effect and react appropriately.

There are people who would argue that the craftsmanship of data management on its own is enough, but it is not. Basic data management skills will only open the gateway to data handling. Once the gate has opened, you have to know which methodology to apply and which tools to use depending on the business context.

If your background is pure data management; it would be advisable to spend three to five years in clinical drug development to get a foundational background. If you come from a clinical background, you should start reading, programming, and learning about how data is used.

If you do not have either of these skill sets, you should prepare for a long learning curve or change careers. Without having a foot in each camp, you will never be able to successfully manage data. Data cannot lie, and you will ultimately be held accountable.

Stage 2 ensure data quality

Data quality is the foundation for reliable decision making. However, information management is never static. This means ensuring data quality is an ongoing process achieved by managing the data and those processes that create the data. While it may sound simple, it is complicated by the fact that any research and development process looks for answers to questions that have been asked by humans in order to gain answers from other humans. All information transition carries the potential for misinterpretation and translation. To secure data quality, you need to be sure that the meaning of the message has not changed.

IT compensates for the increase in data volumes, and it can be tempting to be seduced by its orderly graphs and reports. But you should never forget your critical sense and common sense. You need to continuously question the data. Is this data a true reflection of the processes and conditions under which it was captured? Is this what was meant by the message of the data?

The sources of quality lie within understanding the specific business process and being able to make associations from the business to the data. Simple techniques like data triangulation, visualization, outliers detection, and medical conditional analysis can help you validate the data and ensure no significant discrepancies are included. This type of validation takes time and effort, but there is no way around it if you are serious about ensuring data quality.

Stage 3 track data

As discussed earlier, total predictability in the clinical trial process is simply not possible, and therefore, data cannot just be taken at face level. Your organization's competitive advantage lies in the ability to handle data and convert it into information.

One solution is to start thinking of the data tracking process as a separate process. This means that the tracking in itself is the goal, not data management or data quality. Data tracking can easily be compared to traditional logistics. Just as logistics might include air, sea, road, or rail transport of goods, information can be submitted in the form of a paper entry, ePRO, remote data entry, device specific (e.g., spirometry), or a hematology parameter in acute care ambulance, and many other sources.

That is why it is important to stay focused upon where the data is in the chain. You should be able to explain where each data element is in the supply chain, where it came from, where it will go next, and the different time stages each data element will go through. The next step is to actively use this information to make decisions related to the two previously described steps. Data always tells a story, so if it is delayed, the reason for the delay may impact other data elements. Remember that when a data element is transported from one system to another, the probability of the data element remaining unchanged is only 50%. Either it is correct or it is not correct. That is why during data management activities, it is necessary to secure data quality via systems validation and verification in order to increase the probability of data elements being correct, close to 100%.

For this reason, data tracking is not only a good service for your information customers, but also a good data risk indicator. During my career as information manager, I have seen some obscene systems and application landscapes handling clinical data, where data is exchanged 5—10 times though independent systems/applications, often by independent departments. The results of these complex systems were distorted data, which was used to make bad decisions that ended up costing money, time, and trust.

System landscapes need to reflect business needs and incorporate tracking at every stage. Decisions are continuously being made at many different points in the information chain. By understanding and continuously reevaluating data, you minimize risk and increase your chance of clinical trial success.

Stage 4 develop the right metrics

Metrics are more than just monitoring of an algorithm. When developing metrics that can have a significant influence on an organization, you need to take into account human behavior and the dynamic nature of your organization. This is summed up neatly in the principle of Le Chatelier, which states that any change in a system at equilibrium results in a shift of the equilibrium in the direction that minimizes the change.

A seesaw is the easiest way to illustrate this principle. As any child can tell you, when the person at the lower end of the seesaw gets off, the equilibrium changes, and the other person comes down. Or if we apply it to information management, what was true before your own deliverable of information to the business created a reaction may no longer be true as the reaction changes reality.

Any given metric will influence the behavior and decisions being taken—for better and for worse. So, consider carefully before firing off a lot of lists, numbers, and visualization of the past activities and estimates of the future, maybe enriched with real-time analysis. Remember that what you are showing is an extract of reality, based upon your judgment and selective skills. Each metric draws on a limited amount of information originating from a gigantic pool of data. While what you show is perfectly true, it is not representative of all data and can easily mislead.

On the other hand, it is just as important that you do not drown your customers in information as you try to show all the good data you can get out of the system. The true master of information management is revealed though what you have not included.

If your customers own more than three metrics for each evaluation scheme, the data owns them. Three metrics is the cutoff point where the customer stops thinking. Instead of using the information as a guide, people let the information be the cause of reaction and stop thinking. Faced with too many metrics, they will stop being critical and instead become addicted to data.

The market of information you are offering should not be replacement of solid knowledge and common sense; it should be a service. Customers should not be blindly following your offerings as their guiding star. Metrics should only ever be a supporting service.

Before you start creating metrics, consider the following 14 questions:

1. Will the consequences of the metrics be changes in the organization that are wanted?
2. Do the people in question like to be monitored?
3. Will the metrics undermine the practice of good cooperation?

4. Is the assumption of causality (cause and effect) correct?
5. How will the metrics be implemented?
6. Who will have insight into the metrics' output?
7. Will the metrics be linked to employee's goals?
8. Should any external stakeholders be involved?
9. Where is the ownership placed for the metrics?
10. Will the metric give additional insights which lead to a broader understanding of how the company reacts?
11. How can we capture feedback?
12. What is the frequency of need changes?
13. How do we handle any dysfunctions that are caused by a metric?
14. How do we capture learning and establish operational development? This last question is covered in more detail in the next step.

Stage 5 operational development: use your knowledge

If you have been through the first four stages, you should now understand the business context around your data; be aware of the need for quality data, the benefits of tracking, and you should have the right metrics to support decision making. The next stage is to use the knowledge you have gained in a knowledge management (KM) system.

This is where technology that can leverage the development of knowledge comes in. To create an optimal knowledge management system, you need to consider all the previously outlined parameters and be highly systematic in the actions you take, so you can use the knowledge you have gained during other stages. Within the system, you need to implement procedures for how to capture the evaluation of metrics, learning, feedback, organization reaction, and operational causality.

The capture of ideas that come from interpretation of the metrics is of particular benefit. The following categories are examples of categories that could be used:

1. Human reaction
2. Operational parameters
3. Logistical perspectives
4. Decisions support
5. Organizational level
6. Data sources identification
7. SQL source code
8. Visuals
9. Your own support unit structures

Once you have got the information into the system, you need to decide who can get the information out. If your KM system does not enable knowledge sharing, it can easily end up being just another database that nobody uses. Consider if it is relevant for knowledge to be shared between departments, between functions, between companies in the same field, or between all companies.

If you allow too many people to share information, it could backfire, resulting in data overload. Alternatively, you might find yourself in a situation where people are reluctant to let their activities be monitored. The recipient of your information must be competent and willing to receive your information. A clear structure on information exchange with information submitted to the right people in the right format will save you considerable time and frustration further down the line.

Stage 6 information supply chain management (ISCM)

ISCM is built on the foundation of the first five stages. It ensures you get the right information to the right recipient at the right time.

1. The right information has to be the right data in the right context, quality, and form.
2. The right time is exactly when you or your customers need it, not too early, not too late.
3. The right recipient is identified though your content and operational knowledge.

To measure the effect of the information, you need to create feedback procedures. This is not only for feedback, such as what do people think of it, but feedback on a more basic level: Did your customer receive the information or not? Did it get there at the right time or not? Did the package meet expectations? Remember, you are not measuring whether the effect is good or bad (i.e., wanted or unwanted) but effect and reaction. In essence, you are measuring if steps one through five were efficient and effective.

There is a full toolbox to help you with ISCM: the established theory of supply chain management. This contains a number of feedback procedures and communications tools, like the bullwhip effect, nonlinear feedback, parallel processing, Just-In-Time theory, supplier-buyer psychology, and more.

Note that ISCM (Information Supply Chain Management) is not the same as tracking described in step three. Tracking focuses on pure information logistics (IL) and does not include any information about data quality nor any information content evaluation. ISCM is taking all perspectives into consideration, like technology, time, service, people, reaction, effect, development, and summing up all the preceding steps to a deliverable to your information customers.

Stage 7 integration of external service suppliers

One of the big changes the Internet has brought is the ability to exchange complex information fast and make it easy and intuitively understandable. This has enabled suppliers to specialize in different stages of the production process.

For example, in the manufacturing industry, this has facilitated the physical replacement of production facilities. Car generators are now produced in one country for almost all automobile brands; windshields are produced in a specialized factory in all different shapes and forms. The catalyst for all of this happening is not the location of the production facility but the ability to exchange and understand the blueprint for the activity.

The same changes are also taking place in R&D teams in the pharmaceutical industry—but with extra communication challenges. Pharmaceutical R&D is like one big dish of spaghetti with lots of interconnected processes. Some are executed in parallel and some serially. With the growth in cross-functionality, processes are no longer linear, and communication lines can easily get tangled. This is compounded by the fact that when you outsource tasks, functions, or processes, you add an extra layer of communication. But the effect is not just an extra layer of communication. It is more likely that it is x2, where x is the number of information transactions taking place between the CRO and business owner.

A prerequisite for ensuring successful cooperation is expectation management and measurement, taking all the previous phases into account. Building up a strong relationship with your suppliers with clearly defined metrics will allow you to build a knowledge base that you can use for any future projects.

This expectation management also applies to the incitements that drive you and your suppliers. Be aware of any changes in incitements during the project. An incitement analysis carried out with your supplier can align your goals and cement the working relationship. A reliable activity which can save you from a lot of problems in connections with outsourcing is the development of an information management strategy.

Stage 8 get it all documented in the COP (also see Fig. 18.1)

The table of content in a Clinical Operations Plan (COP) can vary greatly between companies and between trials, reflecting the indications for which the clinical trials are conducted (Fig. 18.1).

An example of a general COP could contain the elements below:

- Communication plan
- Data flow identification
- IL (information logistics)
- ISCM (information supply chain management)
- DI (data intelligences)
- Data handling plan (DHP), which includes
  - Data sources
  - Data exchange plan
  - Cross-check activities
  - Metrics identification
  - SQL development
- Data quality review committee plan
- Stakeholder identification
- Recruitment plan/strategy
- Training plan
- Risk analysis plan (including risk grid)
- Site management strategy
- Lessons learned
- Potential in- or out-sourcing plan

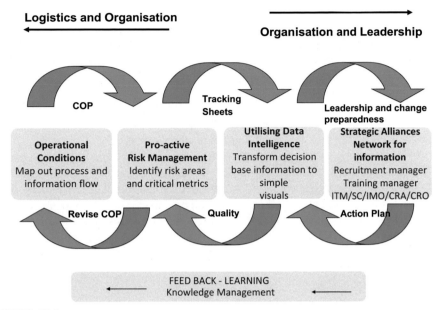

Logistics and Organisation

Organisation and Leadership

COP

Tracking
Sheets

Leadership and change
preparedness

Operational Conditions	Pro-active Risk Management	Utilising Data Intelligence	Strategic Alliances Network for information
Map out process and information flow	Identify risk areas and critical metrics	Transform decision base information to simple visuals	Recruitment manager Training manager ITM/SC/IMO/CRA/CRO

Revise COP

Quality

Action Plan

FEED BACK - LEARNING
Knowledge Management

FIGURE 18.1

The interdependencies of the various activities required for a holistic data management of large volume data, ultimately documented in the Clinical Operations Plan (COP).

SWOT analysis

Strengths: Why Is the New Concept Better? The degree of growth in information and its uncontrolled nature will not be reduced in the future. This approach gives you control without losing vital information. Although it may seem like an unmanageable task, following the eight stages will achieve data quality. The steps are interconnected and contribute to each other, but they can also be executed independently. As the saying goes, the best way to eat an elephant is bit by bit.

Weaknesses: Nothing comes without a cost. The implementation of all eight stages is considerable in time, complexity, and cost. If taken lightly, you can quickly also spend a lot of time and money without obtaining any benefits. This makes it necessary to spend extra time on the structuring of the integrated approach, e.g., through a COP (clinical operation plan).

Opportunities: Ones you go through the eight stages, you will have a solid decisions foundation. It can be reused in other areas in business and can be incorporated as part of the company's culture.

Threats: It takes time. And many companies do not have the ability or money to wait.

Take-home message

Obtaining control over an ever-increasing amount of unstructured data and using this to your advantage is possible. It requires an in-depth knowledge and understanding of the clinical and the technological world, spiced up with some degree of service-mindedness.

The prerequisite for success is to take a structured approach. The first five steps create the foundation; step six creates the exchanges and step seven the integration, with the COP in the last step providing the documentation.

You may nonetheless see systems advertised that offer full data quality and peace of mind, so you can sleep easy at night, safe in the knowledge that your data quality is assured. But these magic solutions do not exist and never will because this sort of software would require total predictability in all processes. With the data growth and the increased frequency of changes we see throughout the industry, the predictability they would require will never be available. The bottom line is that there is no way around hard work; in-depth, well-established knowledge about the business; and good relationships with your customers.

References for further reading

- Guideline for Good Clinical Practice. http://www.ich.org/fileadmin/Public_Web_Site/ICH_Products/Guidelines/Efficacy/E6/E6_R1_Guidlines.pdf.
- European Union law on clinical trials of medicinal products for humans. http://www.ich.org/fileadmin/Public_Web_Site/ICH_Products/Guidelines/Efficacy/E6/E6_R1_Guidlines.pdf.
- US regulations for the conduct of clinical trials. http://www.fda.gov/ScienceResearch/SpecialTopics/RunningClinicalTrials/default.htm.
- EU Guideline on good pharmacovigilance practices (GVP). http://www.ema.europa.eu/docs/en_GB/document_library/Scientific_guideline/2012/06/WC500129138.pdf.
- US FDA Guidance on Quality Risk Management. http://www.fda.gov/downloads/Drugs/GuidanceComplianceRegulatoryInformation/Guidances/UCM073511.pdf.
- ISO. Org: https://www.iso.org/obp/ui/
- The World Bank data: https://data.worldbank.org/indicator/SP.POP.GROW
- European mortality monitoring: https://www.euromomo.eu/
- European Health Information Gateway: https://gateway.euro.who.int/en/hfa-explorer/
- Johns Hopkins University and Medicine Coronavirus Resources Center: https://coronavirus.jhu.edu/us-map.

Share the (digital) knowledge based on quality data

19

Liselotte Hyveled, MscPharm, MMBS

Vice President, Research and Development, Novo Nordisk, Bagsvaerd, Denmark

The need

Over the last 20–30 years, the level of information available for processing has increased dramatically, and we can expect massive further increase after the pandemic. In the 1970s when we were exposed to 500 consumer messages daily, to today up to 5000 [1], we engage in 76, recall 12, and act on 5—the equivalent to 0.1%.

In the pharmaceutical world, there has been a similar increase in data generation. For example, a regulatory file has grown from a mere seven pages in 1952, to 930,000 pages in 2008 and 14 million pages in 2013— the majority of which are generated from clinical trials (Novo Nordisk internal data). The growth in data calls for effective Knowledge Management (KM).

Effective KM is needed as the competitive edge to capture attention. The ability to work with, manage, and communicate all aspects of data is a hugely important skill—not least in the pharmaceutical industry. As a result, companies that combine strong data management skills while shifting into a digital workplace of learning are better positioned to increase productivity and the overall quality of regulatory documentation, including those from clinical trials. This chapter examines how managing, visualizing, communicating, and actively reacting and learning from data throughout a trial can help a company maintain or gain a competitive edge.

Most pharmaceutical companies understand the need to gather data that can generate content. However, if managed inefficiently, the result is data dumping— easily causing a person already inundated with information to become irritated.

Therefore, there is a clear need to give your data context. That can be achieved by deciding which data is most important for ensuring a successful clinical trial and then clearly communicating it, ideally in a visual way. Unlike written reports, data that has been turned into visuals is often clearer to understand and more easily grasped. The onslaught of digital devices (tablets, iPhone apps, chat rooms, etc.) in the last decade demands capabilities to translate information into easily accessible visuals—an excellent communication solution if it can be optimized and understood clearly.

Innovation in Clinical Trial Methodologies. https://doi.org/10.1016/B978-0-12-824490-6.00012-8

However, clearly communicating the right data is just one part of better knowledge sharing and data management; it also needs to be shared within a well-defined knowledge structure. Relatively new players to the data ecosystem are the SoMe platforms, which allow uncontrolled or nonverified data to circulate quickly (see Fig. 19.1).

The consequences of the unstructured growth of unvalidated data are three fold:

1. There is an increase in data-handling investment because there is a greater amount of data to handle.
2. The effectiveness of generating data into useable knowledge is being washed out by the large pool of unstructured unvalidated data.
3. The human bandwidth is limited to absorb a constant amount of input per second; when overloaded, the intake shuts down reducing even good data into noise.

Hence its vital to ensure that knowledge builds on a solid foundation, consisting of understanding the processing causality; that means the sender and receiver of data/knowledge both have the same clear understanding of what the data can show and, especially, what it can't show.

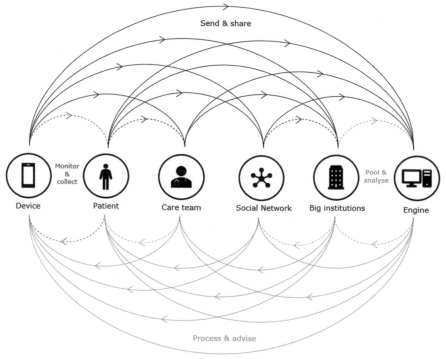

FIGURE 19.1

An example of a data ecosystem in clinical trials, indicating the many sources of data.

As a consequence of the change in methodology during and after the pandemic, we can expect a steep increase in data from Device and Patient (Reported Outcomes) sources.

Trust in data is not limited to mathematics and algorithms. Rather it is based on the origin of the data and built upon the 3Ts:

- Trust (Does the data reflect where it came from?)
- Truth (Is the data telling the true story even when processed?)
- Traceability (Can we always find our way back and forward in the data?)

The solution

This chapter examines how better knowledge sharing and data management throughout a trial enable swift action when needed (Fig. 19.2).

Processes and tools are usually predictable assets whereas people organizations and data tend to be more unpredictable. Data is naturally unpredictable in clinical research; otherwise, we would just have a production process. Hence, we try to control the predictable assets via a lean mindset and understand the unpredictable via an agile mindset. Finding the balance between unpredictable data and predictable processes is key to managing a complex research organization.

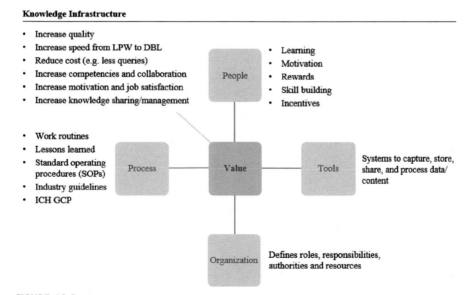

FIGURE 19.2

An optimal knowledge infrastructure for monitoring, sharing, and acting efficiently on data throughout a clinical trial. The structure has four components: tools, process, organization, and people.

Identifying critical data

Once you have identified the elements of your data ecosystem (Fig. 19.1), you need to map and prioritize your knowledge assets. That means creating an inventory list that describes the business needs and key value drivers in the trial. What is needed to accomplish the trial's organizational goals within the optimal speed and quality? In other words, what core knowledge/data must be captured, and what is the difference between what is known and what needs to be known or generated (also see Fig. 19.3)?

The following questions can guide you through the process:

- Which data defines the primary or secondary endpoints in the trial?
- Which data indicates a process flow?
- Are any pieces of data interdependent?
- What is the time frame for collecting the data?
- What are data volume, velocity, variety, and veracity? (See Fig. 19.4)
- What data points are influential on others?
- Where can data be triangulated?
- What data is contradictive? (e.g., How can you have a normal coagulations factor lab sample after the patient died of Disseminated Intravascular Coagulation?)

The data is comparable to circumstantial evidence in a courtroom. In other words, it tells a story.

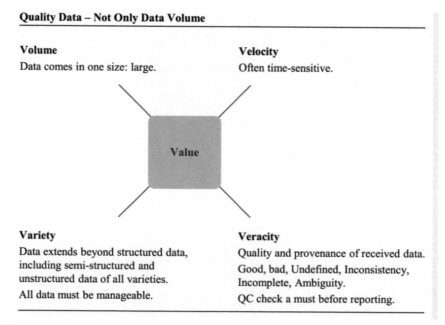

Quality Data – Not Only Data Volume

Volume
Data comes in one size: large.

Velocity
Often time-sensitive.

Value

Variety
Data extends beyond structured data, including semi-structured and unstructured data of all varieties.
All data must be manageable.

Veracity
Quality and provenance of received data.
Good, bad, Undefined, Inconsistency, Incomplete, Ambiguity.
QC check a must before reporting.

FIGURE 19.3

The "4V" which characterize data in a clinical trial. It is useful to remember that quality data is not defined by volume.

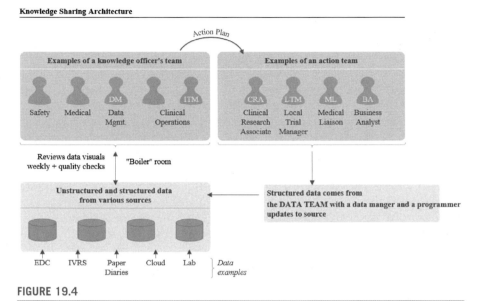

Knowledge Sharing Architecture

FIGURE 19.4

There are three main groups that work together to share relevant knowledge: the data team, the knowledge team, and the action team.

The knowledge structure in practice

Having identified the data you want to capture and generate, you then need to involve the right stakeholders and teams identified for the relevant organizational knowledge (see Fig. 19.4): The data, the knowledge and the action team. These teams work together to continuously collect process and act on the data. Using their combined knowledge and skills, they can accurately highlight any areas of concern and then act swiftly to minimize trial errors and potential audit findings. The cycle of data and knowledge-sharing creates continuous improvement throughout the life cycle of the clinical trial. Let us take a closer look at the separate groups involved.

With the solid anchoring of the digital technology (being defined as devices and systems that use digital or computerized methods to transmit data), the teams detailed in Fig. 19.4 can interact within a virtual workspace to enhance learning and job performance. Physical presence of team members is no longer needed as we got used to virtual meetings, shared platforms, or chat-apps.

The data team works closely with the knowledge team to identify the specific trial data required. With the latter in place, the data team then identifies the source and nature of the input and transforms this data into easily comprehensible visuals. That means they are simple, clear, and intuitive—so minimal explanation is needed. Visuals should be consistent throughout the trial—with special visuals requested if necessary. Be prepared for several review rounds with the knowledge team when creating the visuals, but even when they have been finalized, it is recommended that the data team continue to review the format and content regularly to avoid data report "blindness," which can occur with repeated review of the data. To be clear, the data team is responsible for ensuring the before-mentioned 3T perspective (Trust, Truth, Traceability) on data so quality reviews are critical.

Easy access is also key, so the use of simple tailormade apps or shared data platforms for smartphone or tablets is essential.

The knowledge team is a group of critical stakeholders and represents different aspects of the business with their own spheres of strategic knowledge and experience. Their role is to convene regularly, ideally weekly, in the virtual boiler room to review data. They use the visual reports to identify gaps, errors, and breakdowns that have occurred in data collection. At the end of each meeting, they draw up a plan detailing the needed corrective actions. In addition, live cases from chat-apps or team sites are reviewed. In that way, the knowledge officers can highlight any areas of concern that can impact the trial before their consequences are felt further downstream. Quite simply, it is preventive action before the need for a corrective action occurs.

The action team, unlike the knowledge teams which have a more strategic mindset, is typically driving operations in the clinical setting, and hence, is responsible for executing the action plan.

They are in close contact with all stakeholders who have some form of direct or indirect interaction with subjects. Finally, they provide ongoing feedback to the knowledge team, which is instrumental in the review meetings.

The action team uses the action plan to take local action (sometimes via the CRA) and then provides feedback to the team in the boiler room. This creates an ongoing basis for review at the next weekly knowledge sharing meeting—in effect, creating a continuous cycle of improvement throughout the life cycle of the clinical trial.

The boiler room is where the virtual knowledge team meetings are held. It should be designed for short (approximately 30 min), effective meetings via telepresence if the team is spread over several physical locations. At every meeting, each knowledge officer in the team presents selected data visuals or real-time cases for immediate feedback. That results in an action plan of corrective actions. During these meetings a model for problem solving can be applied as a support for generating the action plan:

> Problem: What is the root cause of the problem (casualty and data-based analysis)?
> Cause: Why has the process broken down (Process and human activities with timestamp perspective)?
> Temporary solution: What can we do straight away to prevent the problem escalating?
> Permanent solution: What can we do to permanently prevent the issue?

A prerequisite for success in this model is that the team captain must feed request to the boiler room. If that does not happen, the communications are based on assumptions— and there is nothing worse than assuming that all is good when in reality it is not.

Real-world example

In this example, the clinical trial team (International trial managers, CRA et al.) are all connected in a chat-app; the app has access to the main data base, allowing easy access to patient recruitment numbers globally, but also allows filtering across many values that provide real-time information and allows for immediate action or feedback to the knowledge- and action teams. Also see previous section "Data analytics for remote monitoring" for additional examples for technical solutions.

The knowledge-sharing architecture described above consisted of a data team, a knowledge team, and an action team that had originally been set up alongside with the use of digital technology. The three groups formed a team that worked closely to monitor if the investigators were following the trial-specific process and guidelines to ensure timely interaction with the local trial staff if any corrective action was necessary.

The knowledge officers in the project start out by deciding which key data from the trial must be transformed into data visuals and how they should be communicated (via team sites, apps, etc.).

During the trial, three issues were quickly identified.

1. Huge country variation in recruitment was seen
2. Data was missing in the patient-related outcome questionnaires
3. Low visibility of patient labels at one site, making drug accountability a challenge

As next step, the knowledge team defined the corrective action to be taken during their weekly action plan. That plan was passed on to the action team for operational implementation and subsequently followed up the week after.

(1) The variation in recruitment was due to variation in local hospital procedures and once optimal procedures were shared and implemented globally, the issues were improved.
(2) The missing data was due to a misunderstanding of the question, as it was poorly formulated, and once an action of creating a clarification and sharing this globally was executed the issue was also improved and complete dataset was collected going forward.
(3) Challenges in drug accountability due to low label visibility were quickly determined to be a one-site issue. The root cause was the pharmacist handling the drugs while using an alcohol rub on the products which took away some of the label text. An issue very easy to act promptly on.

One could ask why a corrective action plan is not just carried out by the clinical research associate (CRA) as a routine check. The answer lies in the fact that the CRA often suffers from information overload, due to the ever-increasing demand for data surveillance, data entry, and reporting. Often, they are unable to see the connection between missing data and a breakdown in the entire process. Worse yet, they are often unable to see the consequence of such issues further down the data stream.

That leads us to the important subject of organizational psychology: all the above activities must give meaning to the individual parties in the process.

Organizational psychology

So far, we have looked at why digital technology has the potential to generate fast, quality information in real time for speedy and appropriate action. We have looked at why the data ecosystem in a clinical trial setting is important; and how to create your knowledge-sharing architecture and provided some tangible questions to consider when identifying your data report need and potential choice of digital technology tools. Finally, we examined a real-world example.

Now we need to focus on the human aspect of the process and consider how to ensure the teams are engaged in sharing, understanding, and acting on the data. It is a well-known fact from the self-determination theory [1] that humans are motivated by three key assets; autonomy, competencies, and greater purpose. Of course, there are also the classic incentives such as compliance and motivation in the form of reward. But the true value of sharing knowledge based on quality data comes when the teams are committed to the project via their engagement in the greater purpose (developing drugs to improve lives), utilizing their competencies, and having influence on the actions to improve their deliveries. Here, data is the connecting factor for the network. The curiosity to survey, look, and understand the data captures the desire for attention to this exact space. It creates a sense of community around the trial. Once established, the community's involvement, commitment, creativity, and ownership evolve—leading to greater self-esteem in team members. Knowledge sharing is no longer an autonomic process on a fixed schedule but a master class—a mindset. That takes both time and effort, and requires continuous, clear, and transparent goals. This is organizational psychology in practice.

Looking at the real-world example provided above, one of the most effective ways of creating commitment came from the virtual boiler room meetings. Commitment to these meetings did not happen instantly. It had to be established. In fact, it took approximately 3 months before the meetings were prioritized and valued by most team members, and even longer before the various groups began to work efficiently as a team. That is to be expected: the team has to go through a period of adaptation and experience first-hand the results of this way of working.

It should also be remembered that the knowledge of the team is intangible as well as explicit. Transferring tacit knowledge is more effective through human interaction and therefore the boiler room framework is much more effective than emailing reports to other stakeholders.

Moreover, it also adds to the valuable social side of sharing knowledge.

Due to the speed of the data and knowledge exchange, a network of people with broad tacit and explicit knowledge is needed to keep pace with the changes in the processes or procedures at the investigational sites. As a collaborative team, the impact of their knowledge is highly effective. However, it requires that everyone recognizes the other members as being of equal importance. All team members bring their own set of skills and experience to the meeting and must be able to translate their knowledge into something everyone present can relate to and understand. Egos and self-importance must be left outside the boiler room door as they will hinder this collaboration.

That is why when knowledge is to be shared, it is not enough to ensure two-way communication where you exchange "data and services." The optimal knowledge ecosystem requires personal engagement, face-to-face interaction, and recognition of the importance of each team member. The trust between team members is a must, as is personal integrity and ethics—in particular, to when it comes to interpretation of facts, data, as well as rumors that could be circulating on various social media. Those should always be evaluated and validated before any potential interaction is taken, even if it is tempting to defend your actions and domain—which ultimately should lead us back to the 3T perspective.

SWOT analysis

Strengths: Collaborative approach allows early identification of issues and corrective action that can reduce time spent on drug development and increase competitive advantage. On the intangible side you may also see increased people skills, an increased sense of trust, more openness, and greater curiosity on the part of your team. Instead of blindly following standard operating procedures, they are forced to consider what action should be taken. All of those factors will ultimately increase your company's capacity to innovate.

Weaknesses: New way of working requires commitment and change in mindset from those in the various teams.

Opportunities: Fast and effective insight into the process and identification of potential gaps.

Minimization of errors and audit findings and a reduction in queries. Better quality of pivotal clinical trial data generated for regulatory approvals.

Threats: It takes time and commitment for the process to be effective and bring value.

One needs to spend more time on handling and transforming data due to the higher amount of data available.

Take-home message

When a knowledge-sharing architecture has been implemented based on your individual data ecosystem, the value that it can generate for a business includes:

- An increased competitive edge by ensuring attention from employees and their desire to use time for such organized information to increase own effectiveness and quality of work.
- Create a strong network of key people with a data-sharing process and a virtual space —"the boiler room"— with prioritized focused meeting slots.

- Understand the data and its sources, identification of new optimized ways of working, choose the right visualization and the best digital tools for your organization, avoiding data overload.
- Fast and effective on time insight into the process and identification of potential gaps utilizing new digital opportunities allow for early identification of issues in a clinical trial, corrective actions to be implemented, and changes to be made quickly.
- Identify the data that is critical for trial success and better quality of pivotal clinical trial data generated for regulatory approvals with less errors, audit findings, and a reduction in queries and time spent on drug development.

However, it takes upfront time, every time, for this process to be effective and it takes time for people to connect via data.

Reference
[1] Yankelovich consumer research.

Further reading
[1] Deci EL, Ryan RM. Intrinsic motivation and self-determination in human behavior. New York, NY: Plenum Press; 1985.

Conclusions

Conclusion

Peter Schüler, MD

Senior Vice President, Drug Development Services Neurosciences, ICON, Langen, Germany

The pandemic has forced our most conservative industry to "dare the innovation." The hope is that those methodological changes will persist once the crisis is over—and we will not return to "business as before."

It would be in the industry's interest to continue with more efficient processes as described throughout this textbook, such as the following:

1. More use of data analytics for decision making
2. Improved data and knowledge management
3. More innovative/adaptive study designs
4. Less complex study protocols with more naturalistic (e.g., digital) endpoints

It would also be in our interest to fully implement the concept of "patient centricity." Even though there is the option at the horizon sometime in the future to run studies "in silico," we currently fully depend on the patients and their caregivers.

The fact that only a very minor portion of potentially eligible patients are willing to get involved in most innovative treatment concepts is embarrassing. Even in oncology, participation in clinical trials typically does not exceed 5% of patients [1].

That has many reasons. One of these is the predominant culture that research is performed on patients, not with patients [2].

Even if we accept that fact, we could do much better by designing more patient-friendly studies, e.g., with less site visits, less assessments, and instead more activities at patient's homes, ideally with naturalistically captured endpoints of Daily Activities through people's smartphones.

While performing that overdue paradigm shift, another major gap opens up in our industry: the lack of methodological validation. Research on clinical research is hard to find and rarely peer-reviewed. In other words, our industry that invented the Gold Standard of Randomized Controlled Trials (RCT) does not apply that standard to its own solutions. At best we rely on historic data from previous studies to assess the benefit of a new process. In only very exceptional circumstances authorities would grant marketing authorization based on a such slim level of evidence.

Nonetheless, in 2019 our industry invested globally USD 46.8 billion in clinical trials [3], which were designed based on "common beliefs" instead of sound data.

Innovation in Clinical Trial Methodologies. https://doi.org/10.1016/B978-0-12-824490-6.00018-9

Instead, we should start to appropriately randomize and initially test any new technology in 50% of the study population while the remaining 50% are exposed to the "good old stuff." Only such research will guide us to optimally plan, execute, and report clinical trials and how to spend this money more efficiently.

References

[1] The Patient Access to Cancer Care Excellence (PACE) [homepage on the Internet] Available from: https://pacenetwork.com/ [Ref list]. [Accessed September 28, 2015].

[2] Thornton S. Beyond rhetoric: we need a strategy for patient involvement in the health service. BMJ 2014;348:g4072.

[3] Clinical Trials Market Size, Share & Trends Analysis Report By Phase (Phase I, Phase II, Phase III, Phase IV), By Study Design (Interventional, Observational, Expanded Access), By Indication, And Segment Forecasts, 2020 − 2027; https://www.grandviewresearch.com/industry-analysis/global-clinical-trials-market.

Index

'Note: Page numbers followed by "f" indicate figures and "t" indicate tables.